# A Concise Treatise on Quantum Mechanics in Phase Space

# A Concise Treatise on Quantum Mechanics in Phase Space

Thomas L Curtright
*University of Miami, USA*

David B Fairlie
*University of Durham, UK*

Cosmas K Zachos
*Argonne National Laboratory, USA*

*Published by*

World Scientific Publishing Co. Pte. Ltd.

5 Toh Tuck Link, Singapore 596224

*USA office:* 27 Warren Street, Suite 401-402, Hackensack, NJ 07601

*UK office:* 57 Shelton Street, Covent Garden, London WC2H 9HE

**Library of Congress Cataloging-in-Publication Data**
Curtright, Thomas.
A concise treatise on quantum mechanics in phase space / Thomas L. Curtright, David B. Fairlie, Cosmas K. Zachos.
pages cm
Includes bibliographical references and index.
ISBN 978-9814520430 (hardcover : alk. paper)
1. Quantum theory--Textbooks. 2. Phase space (Statistical physics)--Textbooks. I. Fairlie, David. II. Zachos, Cosmas. III. Title.
QC174.12.C87 2013
530.12--dc23

2013018153

**British Library Cataloguing-in-Publication Data**
A catalogue record for this book is available from the British Library.



Typeset by Stallion Press
Email: enquiries@stallionpress.com

Printed in Singapore by B & Jo Enterprise Pte Ltd

*Je n'ai fait celle-ci plus longue que parce que je n'ai pas eu le loisir de la faire plus courte.*

*B. Pascal*, Lettres Provinciales XVI *(1656)*

# Contents

# Preface

Wigner's quasi-probability distribution function in phase space is a special (Weyl–Wigner) representation of the density matrix. It has been useful in describing transport in quantum optics, nuclear physics, quantum computing, decoherence, and chaos. It is also of importance in signal processing and the mathematics of algebraic deformation. A remarkable aspect of its internal logic, pioneered by Groenewold and Moyal, has only emerged in the last quarter-century: it furnishes a third, alternative, formulation of quantum mechanics, independent of the conventional Hilbert space or path integral formulations.

In this logically complete and self-standing formulation, one need not choose sides between coordinate or momentum space. It works in full phase space, accommodating the uncertainty principle, and it offers unique insights into the classical limit of quantum theory: the variables (observables) in this formulation are c-number functions in phase space instead of operators, with the same interpretation as their classical counterparts, but are composed together in novel algebraic ways.

This treatise provides an introductory overview and includes an extensive bibliography. Still, the bibliography makes no pretense to exhaustiveness. The overview collects often-used practical formulas and simple illustrations, suitable for applications to a broad range of physics problems, as well as teaching. As a concise treatise, it provides supplementary material which may be used for an advanced undergraduate or a beginning graduate course in quantum mechanics. It represents an expansion of a previous overview with selected papers collected by the authors, and includes a historical narrative account that the subject is due. This Historical Survey is presented first, in Section 1, but it may be skipped by students anxious to get to the mathematical details beginning with the Introduction in Section 2. Alternatively, Section 1 alone may be read by anyone interested only in the history of the subject.

Peter Littlewood and Harry Weerts are thanked for allotting time to make the treatise better.

*T. L. Curtright, D. B. Fairlie, and C. K. Zachos*

# 1. Historical Survey

## The Veridical Paradox

When Feynman first unlocked the secrets of the path integral formalism and presented them to the world, he was publicly rebuked[a]: "It was obvious, Bohr said, that such trajectories violated the uncertainty principle."

However, in this case,[b] Bohr was wrong. Today path integrals are universally recognized and widely used as

R. Feynman

N. Bohr

[a] J. Gleick, *Genius*, Pantheon Books (1992), p. 258.
[b] Unlike (http://en.wikipedia.org/wiki/Bohr-Einstein_debates), the more famous cases where Bohr criticized thought experiments proposed by Einstein, at the 1927 and 1930 Solvay Conferences.

an alternative framework to describe quantum behavior, equivalent to although conceptually distinct from the usual Hilbert space framework, and therefore completely in accord with Heisenberg's uncertainty principle. The different points of view offered by the Hilbert space and path integral frameworks combine to provide greater insight and depth of understanding.

Similarly, many physicists hold the conviction that classical-valued position and momentum variables should not be simultaneously employed in any meaningful formula expressing quantum behavior, simply because this would also seem to violate the uncertainty principle (see the comments about Dirac on page 16, et seq.).

However, they too are wrong. Quantum mechanics (QM) *can* be consistently and autonomously formulated in phase space, with c-number position and momentum variables simultaneously placed on an equal footing, in a way that fully respects Heisenberg's principle. This other quantum framework is equivalent to both the Hilbert space approach and the path integral formulation. Quantum mechanics in phase space (QMPS) thereby gives a third point of view which provides still more insight and understanding.

What follows is the somewhat erratic story of this third formulation [CZ12].

**So fasst uns das, was wir nicht fassen konnten, voller Erscheinung... [Rilke]**

The foundations of this remarkable picture of quantum mechanics were laid out by H. Weyl and E. Wigner around 1930.

H. Weyl

W. Heisenberg and E. Wigner

But the full, self-standing theory was put together in a crowning achievement by two unknowns, at the very beginning of their physics careers and independently of each other, during World War II: H. Groenewold in Holland and J. Moyal in England (see the brief biographies for Groenewold, page 22 et seq., and Moyal, page 26 et seq.). It was only published after the end of the war, under not inconsiderable adversity in the face of opposition by established physicists; and it took quite some time for this uncommon achievement to be appreciated and utilized by the community.[c]

The net result is that quantum mechanics works smoothly and consistently in phase space, where position coordinates and momenta blend together closely and symmetrically. Thus, sharing a common arena and language with classical mechanics,[d] QMPS connects to its classical limit more naturally and intuitively than in the other two familiar alternate pictures, namely, the standard formulation through operators in Hilbert space and the path integral formulation.

Still, as every physics undergraduate learns early on, classical phase space is built out of "c-number" position coordinates and momenta, $x$ and $p$, ordinary commuting variables characterizing physical particles, whereas such observables are usually represented in quantum theory by operators that do not commute. How then can the two be reconciled? The ingenious technical solution to this problem was provided by Groenewold in 1946, and consists of a special binary operation, the $\star$-product, which enables $x$

[c] Perhaps this is because it emerged nearly simultaneously with the path integral and associated diagrammatic methods of Feynman, whose flamboyant application of those methods to the field theory problems of the day captured the attention of physicists worldwide, and thus overshadowed other theoretical developments.

[d] D. Nolte, "The tangled tale of phase space" *Phys. Today*, (April 2010), pp. 33–38.

and $p$ to maintain their conventional classical interpretation, but which also permits $x$ and $p$ to combine more subtly than conventional classical variables; in fact to combine in a way that is *equivalent* to the familiar operator algebra of Hilbert space quantum theory.

Nonetheless, expectation values of quantities measured in the lab (observables) are computed in this picture of quantum mechanics by simply taking integrals of conventional functions of $x$ and $p$ with a quasi-probability density in phase space, the Wigner function — essentially the density matrix in this picture. But, unlike a Liouville probability density of classical statistical mechanics, this density can take provocative negative values and, indeed, these can be reconstructed from lab measurements.[e]

How does one interpret these "negative probabilities" in phase space? It turns out that, like a magical invisible mantle, the uncertainty principle manifests itself in this picture in unexpected but quite powerful ways, and prevents the formulation of unphysical questions, let alone paradoxical answers, associated with such negative values.

Today, students of QMPS routinely demonstrate as an exercise that, in $2n$-dimensional phase space, domains where the WF is solidly negative cannot be significantly

[e] D. Leibfried, T. Pfau, and C. Monroe, "Shadows and mirrors: Reconstructing quantum states of atom motion" *Phys. Today* (April 1998), pp. 22–28.

larger than the minimum uncertainty volume, $(\hbar/2)^n$, and are thus not amenable to direct observation — only indirect inference.

Remarkably, the phase space formulation was reached from rather different, indeed apparently unrelated, directions. To the extent this story has a beginning, this may well have been H. Weyl's remarkably rich 1927 paper [Wey27] shortly after the triumphant formulation of conventional QM. This paper introduced the correspondence of phase space functions to "Weyl-ordered" operators in Hilbert space. It relied on a systematic, completely symmetrized ordering scheme of noncommuting operators $\mathfrak{x}$ and $\mathfrak{p}$.

Eventually it would become apparent that this was a mere change of representation. But as expressed in his paper at the time [Wey27], Weyl believed that this map, which now bears his name, is "the" quantization prescription — superior to other prescriptions — the elusive bridge extending classical mechanics to the operators of the broader quantum theory containing it; effectively, then, some extraordinary "right way" to a "correct" quantum theory.

However, Weyl's correspondence *fails* to transform the square of the classical angular momentum to its accepted quantum analog; and therefore it was soon recognized to be an elegant, but not intrinsically special quantization prescription. As physicists slowly became familiar with the existence of different quantum systems sharing a common classical limit, the quest for the right way to quantization was partially mooted.

In 1931, in establishing the essential uniqueness of Schrödinger's representation in Hilbert space, von Neumann utilized the Weyl correspondence as an equivalent abstract representation of the Heisenberg group in the Hilbert space operator formulation. For completeness' sake, ever the curious mathematician's foible, he worked out the analog (isomorph) of operator multiplication in phase space. He thus effectively discovered the convolution rule governing the noncommutative composition of the corresponding phase space functions — an early version of the $\star$-product.

Nevertheless, possibly because he did not use it for anything at the time, von Neumann oddly ignored his own early result on the $\star$-product and just proceeded to postulate correspondence rules between classical and quantum mechanics in his very influential 1932 book on the foundations of QM.[f] In fact, his ardent follower, Groenewold, would use the $\star$-product to show some of the expectations formed by these rules to be untenable 15 years later. But we are getting ahead of the story.

[f] J. von Neumann, *Mathematical Foundations of Quantum Mechanics*, Princeton University Press (1955, 1983).

J. von Neumann

Very soon after von Neumann's paper appeared, in 1932, Eugene Wigner approached the problem from a completely different point of view, in an effort to calculate quantum corrections to classical thermodynamic (Boltzmann) averages. Without connecting it to the Weyl correspondence, Wigner introduced his eponymous function, a distribution which controls quantum-mechanical diffusive flow in phase space, and thus specifies quantum corrections to the Liouville density of classical statistical mechanics.

As Groenewold and Moyal would find out much later, it turns out that this WF maps to the density matrix (up to multiplicative factors of $\hbar$) under the Weyl map. Thus, without expressing awareness of it, Wigner had introduced an explicit illustration of the inverse map to the Weyl map, now known as the Wigner map.

Wigner also noticed the WF would assume negative values, which complicated its conventional interpretation as a probability density function. However — perhaps unlike his sister's husband — in time Wigner grew to appreciate that the negative values of his function were an asset, and not a liability, in ensuring the orthogonality properties of the formulation's building blocks, the "stargenfunctions".

Wigner further worked out the dynamical evolution law of the WF, which exhibited the nonlocal convolution features of $\star$-product operations, and violations of Liouville's theorem. But, perhaps motivated by practical considerations, he did not pursue the formal and physical implications of such operations, at least not at the time. Those and other decisive steps in the formulation were taken independently by two young novices during World War II.

## A Stay Against Confusion

In 1946, based on his wartime PhD thesis work, much of it carried out in hiding, Hip Groenewold published a decisive paper in which he explored the consistency of the classical–quantum correspondences envisioned by von Neumann. His tool was a fully mastered formulation of the Weyl correspondence as an invertible transform, rather than as a consistent quantization rule. The crux of this isomorphism is the celebrated $\star$-product in its modern form.

Use of this product helped Groenewold demonstrate how Poisson brackets contrast crucially to quantum commutators ("Groenewold's Theorem"). In effect, the Wigner map of quantum commutators is a generalization of Poisson brackets, today called Moyal brackets (perhaps unjustifiably, given that Groenewold's work appeared first), which contains Poisson brackets as their classical limit (technically, a Wigner–Inonü Lie-algebra contraction). By way of illustration, Groenewold further worked out the harmonic oscillator WFs. Remarkably, the basic polynomials involved turned out to be those of Laguerre, and not the Hermite polynomials utilized in the standard Schrödinger formulation! Groenewold had crossed over to a different continent.

At the very same time, in England, Joe Moyal was developing effectively the same theory from a yet different point of view, landing at virtually the opposite coast of the same continent. He argued with Dirac on its validity (see page 16, et seq.) and only succeeded in publishing it, much delayed, in 1949. With his strong statistics background, Moyal focused on all expectation values of quantum operator monomials, $\mathfrak{x}^n \mathfrak{p}^m$, symmetrized by Weyl ordering, expectations which are themselves the numerically valued (c-number) building blocks of every quantum observable measurement.

Moyal saw that these expectation values could be generated out of a *classical-valued characteristic function in phase space*, which he only much later identified with the Fourier transform used previously by Wigner. He then appreciated that many familiar operations of standard quantum mechanics could be apparently bypassed. He reassured himself that the uncertainty principle was incorporated in the structure of this characteristic function, and that it indeed constrained expectation values of "incompatible observables." He interpreted subtleties in the diffusion of the probability fluid and the "negative probability" aspects of it, appreciating that negative probability is a microscopic phenomenon.

Less systematically than Groenewold, Moyal also recast the quantum time evolution of the WF through a deformation of the Poisson bracket into the Moyal bracket, and thus opened up the way for a direct study of the semiclassical limit $\hbar \rightarrow 0$ as an asymptotic expansion in powers of $\hbar$ — "direct" in contrast to the methods of taking the limit of large occupation numbers, or of computing expectations of coherent states. The subsequent applications paper of Moyal with the eminent statistician

M. Bartlett

Maurice Bartlett also appeared in 1949, almost simulaneously with Moyal's fundamental general paper. There, Moyal and Bartlett calculate propagators and transition probabilities for oscillators perturbed by time-dependent potentials, to demonstrate the power of the phase space picture.

By 1949, the formulation was complete, although few took note of Moyal's and especially Groenewold's work. And in fact, at the end of the war in 1945, a number of researchers in Paris, such as J. Yvon and J. Bass, were also rediscovering the Weyl correspondence and converging towards the same picture, albeit in smaller, hesitant, discursive, and considerably less explicit steps.

D. Fairlie (upper left) and E. Wigner (1962)

Important additional steps were subsequently carried out by T. Takabayasi (1954), G. Baker (1958, his thesis), D. Fairlie (1964), and R. Kubo (1964). These researchers provided imaginative applications and filled-in the logical autonomy of the picture — the option, in principle, to derive the Hilbert space picture from it, and not vice versa. The completeness and orthogonality structure of the eigenfunctions in standard QM is paralleled, in a delightful shadow-dance, by QMPS $\star$-operations.

**Be not simply good; be good for something. [Thoreau]**

QMPS can obviously shed light on subtle quantization problems as the comparison with classical theories is more

R. Kubo

systematic and natural. Since the variables involved are the same in both classical and quantum cases, the connection to the classical limit as $\hbar \to 0$ is more readily apparent. But beyond this and self-evident pedagogical intuition, what is this alternate formulation of QM and its panoply of satisfying mathematical structures good for?

It is the natural language to describe quantum transport, and to monitor decoherence of macroscopic quantum states in interaction with the environment, a pressing central concern of quantum computing.[g] It can also serve to analyze and quantize physics phenomena unfolding in

[g]J. Preskill, "Battling decoherence: the fault-tolerant quantum computer" *Phys. Today* (June 1999).

an hypothesized *noncommutative spacetime* with various noncommutative geometries.[h] Such phenomena are most naturally described in Groenewold's and Moyal's language.

However, it may be fair to say that, as was true for the path integral formulation during the first few decades of its existence, the best QMPS "killer apps" are yet to come.

[h]R. J. Szabo, "Quantum field theory on noncommutative spaces" *Phys. Rep.* 378 (2003) 207–299.

**Dirac and QMPS**

P. Dirac

A representative, indeed *authoritative*, opinion, dismissing even the suggestion that quantum mechanics can be expressed in terms of classical-valued phase space variables, was expressed by Paul Dirac in a letter to Joe Moyal on 20 April 1945 (see p. 135, [Moy06]). Dirac said, "I think it is obvious that there cannot be any distribution function $F(p,q)$ which would give correctly the mean value of any $f(p,q)\ldots$" He then tried to carefully explain why he thought as he did, by discussing the underpinnings of the uncertainty relation.

However, in this instance, Dirac's opinion was wrong, and unfounded, despite the fact that he must have been thinking about the subject since publishing some preliminary work along these lines many years before [Dir30]. In retrospect, it is Dirac's unusual misreading of the situation that is obvious, rather than the non-existence of $F(p, q)$.

Perhaps the real irony here is that Dirac's brother-in-law, Eugene Wigner, had already constructed such an $F(p, q)$ several years earlier [Wig32]. Moyal eventually learned of Wigner's work and brought it to Dirac's attention in a letter dated 21 August 1945 (see p. 159 [Moy06]).

E. Wigner

P. Dirac

Nevertheless, the historical record strongly suggests that Dirac held fast to his opinion that quantum mechanics could *not* be formulated in terms of classical-valued phase space variables. For example, Dirac made no changes when discussing the von Neumann density operator, $\rho$, on p. 132 in the final edition of his book.[i] Dirac maintained "Its existence is rather surprising in view of the fact that phase space has no meaning in quantum mechanics, there being no possibility of assigning numerical values simultaneously to the $q$'s and $p$'s." This statement completely overlooks the fact that the Wigner function $F(p,q)$ is precisely a realization of $\rho$ in terms of numerical-valued $q$'s and $p$'s.

How could it be, with his unrivaled ability to create elegant theoretical physics, Dirac did *not* seize the opportunity so unmistakably laid before him by Moyal to return to his very first contributions to the theory of quantum mechanics, and examine in greater depth the relation between classical Poisson brackets and quantum commutators? We will probably never know beyond any doubt — yet another sort of uncertainty principle — but we are led to wonder if it had to do with some key features of Moyal's theory at that time. First, in sharp contrast to Dirac's own operator methods, in its initial stages QMPS theory was definitely *not* a pretty formalism! And, as is

[i]P. A. M. Dirac, *The Principles of Quantum Mechanics*, 4th edn., last revised in 1967 (1958).

well known, beauty was one of Dirac's guiding principles in theoretical physics.

Moreover, the logic of the early formalism was not easy to penetrate. It is clear from his correspondence with Moyal that Dirac did not succeed in cutting away the formal undergrowth to clear a precise conceptual path through the theory behind QMPS, or at least not one that *he* was eager to travel again.[j]

P. Dirac (1960s)

[j]Although Dirac did pursue closely related ideas at least once [Dir45], in his contribution to Bohr's festschrift.

One of the main reasons the early formalism was not pleasing to the eye, and nearly impenetrable, may have had to do with another key aspect of Moyal's 1945 theory: two constructs may have been missing. Again, while we cannot be absolutely certain, we *suspect* the star product and the related bracket were both absent from Moyal's theory *at that time*. So far as we can tell, neither of these constructs appears in any of the correspondence between Moyal and Dirac.

In fact, the product itself is not even contained in the published form of Moyal's work that appeared four years later [Moy49], although the antisymmetrized version of the product — the so-called Moyal bracket — is articulated in that work as a generalization of the Poisson bracket,[k] after first being used by Moyal to express the time evolution of $F(p, q; t)$.[l] Even so, we are not aware of any historical evidence that Moyal *specifically* brought his bracket to Dirac's attention.

Thus, we can hardly avoid speculating, had Moyal communicated *only the contents of his single paragraph about the generalized bracket* to Dirac, the latter would have

[k]See Eq. (7.10) and the associated comments in the last paragraph of Sec. 7, p. 106 [Moy49].

[l]See Eq. (7.8). [Moy49] Granted, the equivalent of that equation was already available in [Wig32], but Wigner did *not* make the sweeping generalization offered by Moyal's equation (7.10).

recognized its importance, as well as its beauty, and the discussion between the two men would have acquired an altogether different tone. For, as Dirac wrote to Moyal on 31 October 1945 (see p 160, [Moy06]), "I think your kind of work would be valuable only if you can put it in a very neat form." The Groenewold product and the Moyal bracket do just that.[m]

---

[m]In any case, by then Groenewold had already found the star product, as well as the related bracket, by taking Weyl's and von Neumann's ideas to their logical conclusion, and had it all published [Gro46] in the time between Moyal's and Dirac's last correspondence and the appearance of [Moy49, BM49], wherein discussions with Groenewold are acknowledged by Moyal.

**Hilbrand Johannes Groenewold**

29 June 1910–23 November 1996[n]

H. Groenewold

Hip Groenewold was born in Muntendam, The Netherlands. He studied at the University of Groningen, from which he graduated in physics with subsidiaries in mathematics and mechanics in 1934.

[n]The material presented here contains statements taken from a previously published obituary, N. Hugenholtz, "Hip Groenewold, 29 Juni 1910–23 November 1996", *Nederlands Tijdschrift voor Natuurkunde* **2** (1997) 31.

In that same year, he went of his own accord to Cambridge, drawn by the presence there of the mathematician John von Neumann, who had given a solid mathematical foundation to quantum mechanics with his book *Mathematische Grundlagen der Quantenmechanik.* This period had a decisive influence on Groenewold's scientific thinking. During his entire life, he remained especially interested in the interpretation of quantum mechanics (e.g. some of his ideas are recounted in Saunders *et al.*[o]). It is therefore not surprising that his PhD thesis, which he completed 11 years later, was devoted to this subject [Gro46]. In addition to his revelation of the star product, and associated technical details, Groenewold's achievement in his thesis was to escape the cognitive straitjacket of the mainstream view that the defining difference between classical mechanics and quantum mechanics was the use of c-number functions and operators, respectively. He understood that these were only habits of use and in no way restricted the physics.

Ever since his return from England in 1935 until his permanent appointment as a lecturer of theoretical physics in Groningen in 1951, Groenewold experienced difficulties finding a paid job in physics. He was

[o]S. Saunders, J. Barrett, A. Kent, and D. Wallace, *Many Worlds?*, Oxford University Press (2010).

an assistant to Zernike in Groningen for a few years, then he went to the Kamerlingh Onnes Laboratory in Leiden, and taught at a grammar school in the Hague from 1940 to 1942. There, he met the woman whom he married in 1942. He spent the remaining war years at several locations in the north of the Netherlands. In July 1945, he began work for another two years as an assistant to Zernike. Finally, he worked for four years at the KNMI (Royal Dutch Meteorological Institute) in De Bilt.

During all these years, Groenewold never lost sight of his research. At his suggestion upon completing his PhD thesis, in 1946, Rosenfeld, of the University of Utrecht, became his promoter, rather than Zernike. In 1951, he was offered a position at Groningen in theoretical physics: first as a lecturer, then as a senior lecturer, and finally as a professor in 1955. With his arrival at the University of Groningen, quantum mechanics was introduced into the curriculum.

In 1971 he decided to resign as a professor in theoretical physics in order to accept a position in the Central Interfaculty for teaching Science and Society. However, he remained affiliated with the theoretical institute as an extraordinary professor. In 1975 he retired.

In his younger years, Hip was a passionate puppet player, having brought happiness to many children's hearts with beautiful puppets he made himself. Later, he was especially interested in painting. He personally knew several painters, and owned many of their works. He was a great lover of the after-war CoBrA art. This love gave him much comfort during his last years.

**José Enrique Moyal**

1 October 1910–22 May 1998[p]

J. Moyal

Joe Moyal was born in Jerusalem and spent much of his youth in Palestine. He studied electrical engineering in France, at Grenoble and Paris, in the early 1930s. He then worked as an engineer, later continuing his studies in

[p]The material presented here contains statements taken from a previously published obituary, J. Gani, "Obituary: José Enrique Moyal", *J Appl Probab* **35** (1998) 1012–1017.

mathematics at Cambridge, statistics at the Institut de Statistique, Paris, and theoretical physics at the Institut Henri Poincaré, Paris.

After a period of research on turbulence and diffusion of gases at the French Ministry of Aviation in Paris, he escaped to London at the time of the German invasion in 1940. The eminent writer C. P. Snow, then adviser to the British Civil Service, arranged for him to be allocated to de Havilland's at Hatfield, where he was involved in aircraft research into vibration and electronic instrumentation.

During the war, hoping for a career in theoretical physics, Moyal developed his ideas on the statistical nature of quantum mechanics, initially trying to get Dirac interested in them, in December 1940, but without success. After substantial progress on his own, his poignant and intense scholarly correspondence with Dirac (February 1944 to January 1946, reproduced in [Moy06]) indicates he was not aware, at first, that his phase space statistics-based formulation was actually equivalent to standard QM. Nevertheless, he soon appreciated its alternate beauty and power. In their spirited correspondence, Dirac patiently but insistently recorded his reservations, with mathematically trenchant arguments, although lacking essential appreciation of Moyal's novel point of view: a radical departure from the conventional Hilbert space picture [Moy49]. The correspondence ended in anticipation of a Moyal colloquium at Cambridge in early 1946.

That same year, Moyal's first academic appointment was in Mathematical Physics at Queen's University Belfast. He was later a lecturer and senior lecturer with M. S. Bartlett in the Statistical Laboratory at the University of Manchester, where he honed and applied his version of quantum mechanics [BM49].

In 1958, he became a Reader in the Department of Statistics, Institute of Advanced Studies, Australian National University, for a period of 6 years. There he trained several graduate students, now eminent professors in Australia and the USA. In 1964, he returned to his earlier interest in mathematical physics at the Argonne National Laboratory near Chicago, coming back to Macquarie University as Professor of Mathematics before retiring in 1978.

Joe's interests were broad: he was an engineer who contributed to the understanding of rubber-like materials, a statistician responsible for the early development of the mathematical theory of stochastic processes, a theoretical physicist who discovered the "Moyal bracket" in quantum mechanics, and a mathematician who researched the foundations of quantum field theory. He was one of a rare breed of mathematical scientists working in several fields, to each of which he made fundamental contributions.

# 2. Introduction

There are at least three logically autonomous alternative paths to quantization. The first is the standard one utilizing operators in Hilbert space, developed by Heisenberg, Schrödinger, Dirac, and others in the 1920s. The second one relies on path integrals, and was conceived by Dirac [Dir33] and constructed by Feynman.

The third one (the bronze medal!) is the phase space formulation surveyed in this book. It is based on Wigner's (1932) quasi-distribution function [Wig32] and Weyl's (1927) correspondence [Wey27] between ordinary c-number functions in phase space and quantum-mechanical operators in Hilbert space.

The crucial quantum-mechanical composition structure of all such functions relies on the so-called $\star$-product, as was fully understood by Groenewold (1946) [Gro46], who, together with Moyal (1949) [Moy49], put the entire formulation together. Still, insights on interpretation and a full appreciation of its conceptual autonomy, as well as its distinctive beauty, took some time to mature with the work of Takabayasi [Tak54], Baker [Bak58], and Fairlie [Fai64], among others, as sketched in the preceding Historical Survey section.

This complete formulation is based on the Wigner function (WF), which is a quasi-probability distribution function in phase space,

$$f(x,p) = \frac{1}{2\pi} \int dy \, \psi^* \left( x - \frac{\hbar}{2} y \right) e^{-iyp} \psi \left( x + \frac{\hbar}{2} y \right). \quad (1)$$

It is a generating function for all spatial autocorrelation functions of a given quantum-mechanical wavefunction $\psi(x)$. More importantly, it is a special representation of the density matrix (in the Weyl correspondence, as detailed in Sec. 12).

Alternatively, in a $2n$-dimensional phase space, it amounts to

$$f(x,p) = \frac{1}{(2\pi\hbar)^n} \int d^n y \left\langle x + \frac{y}{2} \right| \boldsymbol{\rho} \left| x - \frac{y}{2} \right\rangle e^{-ip\cdot y/\hbar}, \quad (2)$$

where $\psi(x) = \langle x | \psi \rangle$ in the density operator $\boldsymbol{\rho}$,

$$\boldsymbol{\rho} = \int d^n z \int d^n x d^n p \left| x + \frac{z}{2} \right\rangle f(x,p) \, e^{ip\cdot z/\hbar} \left\langle x - \frac{z}{2} \right| . \quad (3)$$

There are several outstanding reviews on the subject: in [HOS84, Tak89, Ber80, BJ84], [Lit86, deA98, Shi79, Tat83, Coh95, KN91], [Kub64, deG74, KW90, Ber77, Lee95, Dah01, Sch02], [DHS00, CZ83, Gad95, HH02], [Str57, McD88, Leo97, Sny80, Bal75, TKS83, BFF78].

Nevertheless, the central conceit of the present overview is that the above input wavefunctions may ultimately be bypassed, since the WFs are determined, in principle, as the solutions of suitable functional equations in phase space. Connections to the Hilbert space operator formulation of quantum mechanics may thus be ignored, in principle — even though they are provided in Sec. 12 for pedagogy and confirmation of the formulation's equivalence. One might then envision an imaginary world in which this formulation of quantum mechanics had preceded the conventional Hilbert space formulation, and its own techniques and methods had arisen independently, perhaps out of generalizations of classical mechanics and statistical mechanics.

It is not only wavefunctions that are missing in this formulation. Beyond the ubiquitous (noncommutative, associative, pseudodifferential) operation which encodes the entire quantum-mechanical action, the $\star$-product, there are no linear operators. Expectations of observables and transition amplitudes are phase space integrals of c-number functions, weighted by the WF, as in statistical mechanics.

Consequently, even though the WF is not positive semidefinite (it can be, and usually is negative in parts of phase space [Wig32]), the computation of expectations and the associated concepts are evocative of classical probability theory, as emphasized by Moyal. Still, telltale features of quantum mechanics are reflected in the noncommutative multiplication of such c-number phase space functions through the $\star$-product, in systematic analogy to operator multiplication in Hilbert space.

This formulation of quantum mechanics is useful in describing *quantum* hydrodynamic transport processes in phase space, [IZ51] notably in quantum optics [Sch02, Leo97, SM00], nuclear and particle physics [Bak60, Wo82, SP81, WH99, MM84, CC03, BJY04], condensed matter [DO85, MMP94, DBB02, KKFR89] [JG93, BP96, Ram04, KL01, JBM03, Mor09, SLC11], the study of semiclassical limits of mesoscopic systems [Imr67, OR57, Sch69, Ber77, KW87, OM95, MS95, MOT98, Vor89, Vo78], [Hel76, Wer95, Ara95, Mah87, Rob93, CdD04], [Pul06, Zdn06], and the transition to classical statistical mechanics [VMdG61, CL83, JD99, Fre87, SRF03, BD98, Dek77, Raj83], [HY96, CV98, SM00, FLM98, FZ01, Zal03, CKTM07].

Since observables are expressed by essentially *common variables in both their quantum and classical configurations*, this formulation is the natural language in which to investigate quantum signatures of chaos [KB81, HW80, GHSS05, Bra03, MNV08, CSA09, Haa10] and decoherence [Ber77, JN90, Zu91, ZP94, Hab90, BC99, KZZ02, KJ99, FBA96, Kol96, GH93, CL03, BTU93, Mon94, HP03, OC03], [GK94, BC09, GB03, MMM11] (of utility in, e.g. quantum computing [BHP02, MPS02, TGS05]).

It likewise provides suitable intuition in quantum-mechanical interference problems [Wis97, Son09], molecular Talbot–Lau interferometry [NH08], probability flows as negative probability backflows [BM94, FMS00, BV90], and measurements of atomic systems [Smi93], [Dun95, Lei96, KPM97, Lvo01, JS02, BHS02, Ber02, Cas91].

The intriguing mathematical structure of the formulation is of relevance to Lie Algebras [FFZ89], martingales in turbulence [Fan03], and string field theory [BKM03]. It has also been repurposed into M-theory and quantum field theory advances linked to noncommutative geometry [SW99,Fil96] (for reviews, see [Cas00,Har01,DN01,HS02]), and to matrix models [Tay01,KS02], these apply spacetime uncertainty principles [Pei33, Yo89, JY98, SST00] reliant on the $\star$-product. (Transverse spatial dimensions act formally as momenta, and, analogously to quantum mechanics, their uncertainty is increased or decreased inversely to the uncertainty of a given direction.)

As a significant aside, in formal emulation of quantum mechanics [Vill48], the WF has extensive practical applications in signal processing, filtering, and engineering (time-frequency analysis), since, mathematically, time and frequency constitute a pair of Fourier-conjugate variables, just like the position and momentum pair of phase space.

Thus, time-varying signals are best represented in a WF as time-varying *spectrograms*, analogously to a music score: i.e. the changing distribution of frequencies is monitored in time [deB67, BBL80, Wok97, QC96, MH97, Coh95, Gro01, Fla99]: even though the description is constrained and redundant, it furnishes an intuitive picture of the signal which a mere time profile or frequency spectrogram fails to convey.

Applications abound [CGB91, Lou96, MH97] in bioengineering, acoustics, speech analysis, vision processing, radar imaging, turbulence microstructure analysis, seismic imaging [WL10], and the monitoring of internal combustion engine-knocking, failing helicopter-component vibrations, atmospheric radio occultations [GLL10] and so on.

For simplicity, the formulation will be mostly illustrated here for one coordinate and its conjugate momentum, but generalization to arbitrary-sized phase spaces is straightforward [Bal75, DM86], including infinite-dimensional ones, namely scalar field theory [Dit90, Les84, Na97, CZ99, CPP01, MM94]: the respective WFs are simple products of single-particle WFs.

# 3. The Wigner Function

As already indicated, the quasi-probability measure in phase space is the WF,

$$f(x,p) = \frac{1}{2\pi} \int dy\, \psi^* \left( x - \frac{\hbar}{2} y \right) e^{-iyp} \psi \left( x + \frac{\hbar}{2} y \right). \quad (4)$$

It is obviously normalized, $\int dpdx f(x,p) = 1$, for normalized input wavefunctions. In the classical limit, $\hbar \to 0$, it would reduce to the probability density in coordinate space, $x$, usually highly localized, multiplied by $\delta$-functions in momentum: in phase space, the classical limit is "spiky" and certain!

This expression has more $x-p$ symmetry than is apparent, as Fourier transformation to momentum-space wavefunctions, $\phi(p) = \int dx \exp(-ixp/\hbar)\psi(x)/\sqrt{2\pi\hbar}$, yields a completely symmetric expression with the roles of $x$ and $p$ reversed; and, upon rescaling of the arguments $x$ and $p$, a symmetric classical limit.

The WF is also manifestly real.[q] It is further constrained [Bak58] by the Cauchy–Schwarz inequality to be

[q]In one space dimension, by virtue of non-degeneracy, $\psi$ has the same effect as $\psi^*$, and $f$ turns out to be $p$-even; but this is not a property used here.

bounded: $-\frac{2}{\hbar} \leq f(x,p) \leq \frac{2}{\hbar}$. Again, this bound disappears in the spiky classical limit. Thus, this quantum-mechanical bound precludes a WF which is a perfectly localized delta function in $x$ and $p$ — the uncertainty principle.

Respectively, $p$- or $x$-projection leads to marginal probability densities: a space-like shadow $\int dp f(x,p) = \rho(x)$, *or else* a momentum-space shadow $\int dx f(x,p) = \sigma(p)$. Either is a bona fide probability density, being positive semidefinite. But these potentialities are actually interwoven. Neither can be conditioned on the other, as the uncertainty principle is fighting back: the WF $f(x,p)$ itself can, and most often is, *negative* in some *small* areas of phase space [Wig32, HOS84, MLD86]. This is illustrated below, and furnishes a hallmark of QM interference in this language. Such negative features thus serve to monitor quantum coherence; while their attenuation monitors its loss. (In fact, the only pure state WF which is non-negative is the Gaussian [Hud74], a state of maximum entropy [Raj83].)

The counter-intuitive "negative probability" aspects of this quasi-probability distribution have been explored and interpreted [Bar45, Fey87, BM94, MLD86] (for a popular review, see [LPM98]). For instance, negative probability flows may be regarded as legitimate probability backflows in interesting settings [BM94]. Nevertheless, the WF for atomic systems can still be measured in the laboratory, albeit indirectly, and reconstructed [Smi93, Dun95, Lei96, KPM97, Lvo01, Lut96, BAD96, BHS02, Ber02, BRWK99, Vog89].

Smoothing $f$ by a filter of size larger than $\hbar$ (e.g. convolving with a phase space Gaussian, so a Weierstrass transform) necessarily results in a *positive semidefinite function*, i.e. it may be thought to have been smeared, "regularized", or blurred to a classical[r] distribution [deB67, Car76, Ste80, OW81, Raj83].

It is thus evident that phase space patches of uniformly negative value for $f$ cannot be larger than a few $\hbar$, since,

---

[r]This one is called the Husimi distribution [Tak89, TA99], and sometimes information scientists examine it preferentially on account of its non-negative feature. Nevertheless, it comes with a substantially heavy price, as it needs to be "dressed" back to the WF, for all practical purposes, when equivalent quantum expectation values are computed with it: i.e. unlike the WF, it does *not* serve as an immediate quasi-probability distribution with no further measure (see Sec. 13). The negative feature of the WF is, in the last analysis, an asset, and not a liability, and provides an efficient description of "beats" [BBL80, Wok97, QC96, MH97, Coh95], cf. Fig. 1.

*A point of caution:* if, instead, strictly *inequivalent* expectation values were taken with the Husimi distribution *without* the requisite dressing of Sec. 13, i.e. improperly, as though it were a bona fide probability distribution, such expectation values would actually reflect *loss of quantum information*: they would represent semiclassically smeared observables [WO87].

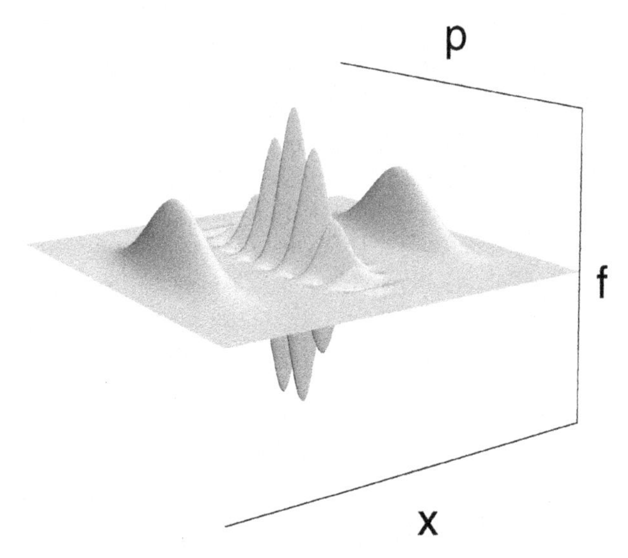


Fig. 1. Wigner function of a pair of Gaussian wavepackets, centered at $x = \pm a$, $f(x, p; a) = \exp(-(x^2 + p^2))(\exp(-a^2)\cosh(2ax) + \cos(2pa))/(\pi(1 + e^{-a^2}))$. (Here, for simplicity, we scale to $\hbar = 1$. The corresponding wavefunction is $\psi(x; a) = (\exp(-(x + a)^2/2) + \exp(-(x - a)^2/2))/(\pi^{1/4}\sqrt{2 + 2e^{-a^2}})$. In this figure, $a = 6$ is chosen, significantly larger than the width of the Gaussians.) Note the phase space interference structure ("beats") with negative values in the $x$ region between the two packets where there is no wavefunction support — hence vanishing probability for the presence of the particle. The oscillation frequency in the $p$-direction is $a/\pi$. Thus, it increases with growing separation $a$, ultimately smearing away the interference structure.

otherwise, smoothing by such an $\hbar$-filter would fail to obliterate them as required above. That is, *negative patches are small, a microscopic phenomenon* in general, in some sense shielded by the uncertainty principle. Monitoring negative WF features and their attenuation in time (as quantum information leaks into the environment) affords a measure of decoherence and drift towards a classical (mixed) state [KJ99].

Among real functions, the WFs comprise a rather small, highly constrained, set. When is a real function $f(x,p)$ a bona fide, pure state, Wigner function of the form (4)? Evidently, when its Fourier transform (the cross-spectral density) "left–right" factorizes,

$$\tilde{f}(x,y) = \int dp\, e^{ipy} f(x,p) = g_L^*(x - \hbar y/2) g_R(x + \hbar y/2). \tag{5}$$

That is,

$$\frac{\partial^2 \ln \tilde{f}}{\partial(x - \hbar y/2)\partial(x + \hbar y/2)} = 0, \tag{6}$$

so that, for real $f$, $g_L = g_R$. An equivalent test for pure states will be given in Eq. (25).

Nevertheless, as indicated, the WF *is* a distribution function, after all: it provides the integration measure in phase space to yield expectation values of observables from corresponding phase space c-number functions. Such functions are often familiar classical quantities; but, in general, they are uniquely associated to suitably ordered operators through *Weyl's correspondence rule* [Wey27].

Given an operator (in gothic script) ordered in this prescription,

$$\mathfrak{G}(\mathfrak{x},\mathfrak{p}) = \frac{1}{(2\pi)^2}\int d\tau d\sigma dx dp\, g(x,p)\exp(i\tau(\mathfrak{p}-p) + i\sigma(\mathfrak{x}-x)), \tag{7}$$

the corresponding phase space function $g(x,p)$ (the *Weyl kernel function*, or the *Wigner transform* of that operator) is obtained by

$$\mathfrak{p} \longmapsto p, \quad \mathfrak{x} \longmapsto x. \tag{8}$$

That operator's expectation value is then given by a "phase space average" [Gro46, Moy49, Bas48],

$$\langle \mathfrak{G} \rangle = \int dx dp\, f(x,p)\, g(x,p). \tag{9}$$

The kernel function $g(x,p)$ is often the unmodified classical observable expression, such as a conventional Hamiltonian, $H = p^2/2m + V(x)$, i.e. the transition from classical mechanics is straightforward ("quantization").

However, the kernel function contains $\hbar$ corrections when there are quantum-mechanical ordering ambiguities in the observables, such as in the kernel of the square of the angular momentum, $\mathfrak{L}\cdot\mathfrak{L}$. This one contains an additional term $-3\hbar^2/2$ introduced by the Weyl ordering

[She59, DS82, DS02], beyond the mere classical expression, $L^2$. In fact, with suitable averaging, this quantum offset accounts for the nontrivial angular momentum $L = \hbar$ of the ground-state Bohr orbit, when the standard hydrogen quantum ground state has vanishing $\langle \mathfrak{L} \cdot \mathfrak{L} \rangle = 0$.

In such cases (including momentum-dependent potentials), even nontrivial $O(\hbar)$ quantum corrections in the phase space kernel functions (which characterize different operator orderings) can be produced efficiently without direct, cumbersome consideration of operators [CZ02, Hie84]. More detailed discussion of the Weyl and alternate correspondence maps is provided in Secs. 12 and 13.

In this sense, expectation values of the physical observables specified by kernel functions $g(x,p)$ are computed through integration with the WF, $f(x,p)$, in close analogy with classical probability theory, despite the non-positive-definiteness of the distribution function. This operation corresponds to tracing an operator with the density matrix (cf. Sec. 12).

**Exercise 1.** When does a WF vanish? To see where the WF $f(x_0, p_0)$ vanishes or not, for a given wavefunction $\psi(x)$ with *bounded support* (i.e. vanishing outside a finite region in $x$), pick a point $x_0$ and reflect $\psi(x) = \psi(x_0 + (x - x_0))$ across $x_0$ to $\psi(x_0 - (x - x_0)) = \psi(2x_0 - x)$. See if the overlap between these two distributions is nontrivial or not, to get $f(x_0, p) \neq 0$ or $= 0$.

Now consider the *schematic* (unrealistic) real $\psi(x)$:

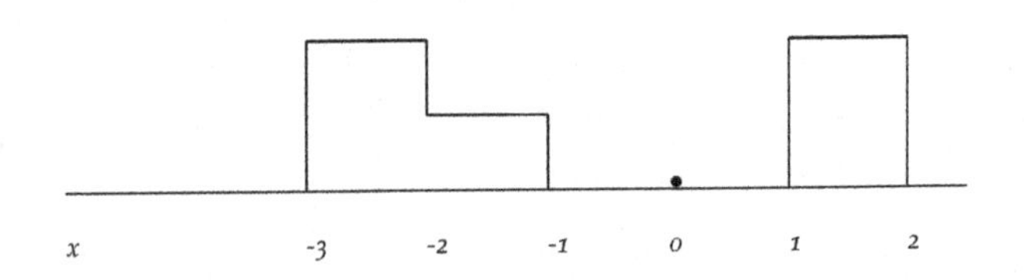


Is $f(x_0 = -2, p) = 0$? Is $f(x_0 = 3, p) = 0$? Is $f(x_0 = 0, p) = 0$?

**Exercise 2.** Consider a particle free to move inside a one-dimensional box of width $a$ with impenetrable ends. The particle is in the ground state given by

$$\psi(x) = \begin{cases} 0 & \text{if} \quad |x| \geq a/2 \\ \sqrt{\frac{2}{a}} \cos(\pi x/a) & \text{if} \quad |x| \leq a/2 \end{cases}$$

Compute the Wigner function $f(x, p)$ for this state.

# 4. Solving for the Wigner Function

Given a specification of observables, the next step is to find the relevant WF for a given Hamiltonian. Can this be done without solving for the Schrödinger wavefunctions $\psi$, i.e. not using Schrödinger's equation directly? Indeed, the functional equations which $f$ satisfies completely determine it.

Firstly, its dynamical evolution is specified by *Moyal's equation.* This is the extension of Liouville's theorem of classical mechanics for a classical Hamiltonian $H(x,p)$, namely $\partial_t f + \{f, H\} = 0$, to quantum mechanics, in this language [Wig32, Bas48, Moy49]:

$$\frac{\partial f}{\partial t} = \frac{H \star f - f \star H}{i\hbar} \equiv \{\{H, f\}\}, \tag{10}$$

where the $\star$-product [Gro46] is

$$\star \equiv e^{\frac{i\hbar}{2}(\overleftarrow{\partial}_x \overrightarrow{\partial}_p - \overleftarrow{\partial}_p \overrightarrow{\partial}_x)}. \tag{11}$$

The right-hand side of (10) is dubbed the "Moyal Bracket" (MB), and the quantum commutator is its Weyl-correspondent (its Weyl transform). It is the essentially unique one-parameter ($\hbar$) associative deformation (expansion) of the Poisson Brackets (PB) of classical mechanics [Vey75, BFF78, FLS76, Ar83, Fle90, deW83, BCG97, TD97]. Expansion in $\hbar$ around 0 reveals that it consists of the Poisson Bracket corrected by terms $O(\hbar)$.

Moyal's evolution equation (10) also evokes Heisenberg's equation of motion for operators (with the suitable sign of von Neumann's evolution equation for the density matrix), except $H$ and $f$ here are ordinary "classical" phase space functions, and it is the $\star$-product which now enforces noncommutativity. This language, then, makes the link between quantum commutators and Poisson Brackets more transparent.

Since the $\star$-product involves exponentials of derivative operators, it may be evaluated in practice through translation of function arguments ("Bopp shifts"),

**Lemma 1.**

$$f(x,p) \star g(x,p) = f\left(x + \frac{i\hbar}{2}\vec{\partial}_p, p - \frac{i\hbar}{2}\vec{\partial}_x\right) g(x,p). \tag{12}$$

The equivalent Fourier representation of the $\star$-product is the generalized convolution [Neu31, Bak58]

$$f \star g = \frac{1}{\hbar^2\pi^2}\int dp'dp''dx'dx''\, f(x',p')\, g(x'',p'') \times \exp\left(\frac{-2i}{\hbar}(p(x'-x'') + p'(x''-x) + p''(x-x'))\right). \tag{13}$$

An alternate integral representation of this product is [HOS84]

$$f \star g = (\hbar\pi)^{-2} \int dp' dp'' dx' dx''\, f(x + x', p + p') \times g(x + x'', p + p'') \exp\left(\frac{2i}{\hbar}(x'p'' - x''p')\right), \quad (14)$$

which readily displays noncommutativity and associativity.

The fundamental Theorem (proof later, see page 96) dictates that $\star$-multiplication of c-number phase space functions is in complete isomorphism to Hilbert space operator multiplication [Gro46] of the respective Weyl transforms,

$$\mathfrak{A}(\mathfrak{x}, \mathfrak{p})\mathfrak{B}(\mathfrak{x}, \mathfrak{p}) = \frac{1}{(2\pi)^2} \int d\tau d\sigma dx dp (a \star b) \times \exp(i\tau(\mathfrak{p} - p) + i\sigma(\mathfrak{x} - x)). \quad (15)$$

The cyclic phase space trace is directly seen in the representation (14) to reduce to a plain product if there is *only one* $\star$ involved,

**Lemma 2.**

$$\int dp dx\, f \star g = \int dp dx\, fg = \int dp dx\, g \star f. \quad (16)$$

Moyal's equation is necessary, but not sufficient to specify the WF for a system. In the conventional formulation of quantum mechanics, systematic solution of time-dependent equations is usually predicated on the spectrum of stationary ones. Time-independent pure state Wigner functions $\star$-commute with $H$; but, clearly, not every function $\star$-commuting with $H$ can be a bona fide WF (e.g. any $\star$-function of $H$ will $\star$-commute with $H$).

Static WFs obey even more powerful functional $\star$-genvalue (pronounced "stargenvalue") equations [Fai64] (also see [Bas48, Kun67, Coh76, Dah83]),

$$\begin{aligned} H(x,p) \star f(x,p) &= H\left(x + \frac{i\hbar}{2}\vec{\partial}_p, p - \frac{i\hbar}{2}\vec{\partial}_x\right) f(x,p) \\ &= f(x,p) \star H(x,p) = E\, f(x,p), \end{aligned} \tag{17}$$

where $E$ is the energy eigenvalue of $\mathfrak{H}\psi = E\psi$ in Hilbert space. These amount to a complete characterization of the WFs [CFZ98]. (NB. Observe the $\hbar \to 0$ transition to the classical limit.)

**Lemma 3.** *For real functions $f(x,p)$, the Wigner form (4) for pure static eigenstates is equivalent to compliance with the $\star$-genvalue equations (17) ($\Re$ and $\Im$ parts).*

**Proof.**

$$\begin{aligned} H(x,p) \star f(x,p) &= \frac{1}{2\pi}\left(\left(p - i\frac{\hbar}{2}\vec{\partial}_x\right)^2 \Big/ 2m + V(x)\right) \\ &\quad \times \int dy\, e^{-iy(p + i\frac{\hbar}{2}\overleftarrow{\partial}_x)} \psi^*\left(x - \frac{\hbar}{2}y\right) \psi\left(x + \frac{\hbar}{2}y\right) \end{aligned}$$

$$= \frac{1}{2\pi} \int dy \left( \left( p - i\frac{\hbar}{2} \overrightarrow{\partial}_x \right)^2 \Big/ 2m + V\left( x + \frac{\hbar}{2} y \right) \right)$$

$$\times e^{-iyp} \psi^* \left( x - \frac{\hbar}{2} y \right) \psi \left( x + \frac{\hbar}{2} y \right)$$

$$= \frac{1}{2\pi} \int dy \, e^{-iyp} \left( \left( i \overrightarrow{\partial}_y + i\frac{\hbar}{2} \overrightarrow{\partial}_x \right)^2 \Big/ 2m + V\left( x + \frac{\hbar}{2} y \right) \right)$$

$$\times \psi^* \left( x - \frac{\hbar}{2} y \right) \psi \left( x + \frac{\hbar}{2} y \right)$$

$$= \frac{1}{2\pi} \int dy \, e^{-iyp} \psi^* \left( x - \frac{\hbar}{2} y \right) E \, \psi \left( x + \frac{\hbar}{2} y \right)$$

$$= E \, f(x,p). \tag{18}$$

Action of the effective differential operators on $\psi^*$ turns out to be null.

Symmetrically,

$$f \star H = \frac{1}{2\pi} \int dy \, e^{-iyp} \left( -\frac{1}{2m} \left( \overrightarrow{\partial}_y - \frac{\hbar}{2} \overrightarrow{\partial}_x \right)^2 + V\left( x - \frac{\hbar}{2} y \right) \right)$$

$$\times \psi^* \left( x - \frac{\hbar}{2} y \right) \psi \left( x + \frac{\hbar}{2} y \right)$$

$$= E \, f(x,p), \tag{19}$$

where the action on $\psi$ is now trivial.

Conversely, the pair of $\star$-eigenvalue equations dictate, for $f(x,p) = \int dy\, e^{-iyp} \tilde{f}(x,y)$,

$$\int dy\, e^{-iyp} \left( -\frac{1}{2m} \left( \overrightarrow{\partial}_y \pm \frac{\hbar}{2} \overrightarrow{\partial}_x \right)^2 + V\left( x \pm \frac{\hbar}{2} y \right) - E \right) \tilde{f}(x,y) = 0. \tag{20}$$

Hence, real solutions of (17) must be of the form

$$f = \int dy\, e^{-iyp} \psi^* \left( x - \frac{\hbar}{2} y \right) \psi \left( x + \frac{\hbar}{2} y \right) / 2\pi,$$

such that $\mathfrak{H}\psi = E\psi$.

Equations (17) lead to spectral properties for WFs [Fai64, CFZ98], as in the Hilbert space formulation. For instance, projective orthogonality of the $\star$-genfunctions (pronounced "stargenfunctions") follows from associativity, which allows evaluation in two alternate groupings:

$$f \star H \star g = E_f\, f \star g = E_g\, f \star g. \tag{21}$$

Thus, for $E_g \neq E_f$, it is necessary that

$$f \star g = 0. \tag{22}$$

Moreover, precluding degeneracy (which can be treated separately), choosing $f = g$ above yields,

$$f \star H \star f = E_f\ f \star f = H \star f \star f, \tag{23}$$

$\star$-genfunction and hence $f \star f$ must be the stargenfunction in question,

$$f \star f \propto f. \tag{24}$$

Pure state $f$s then $\star$-project onto their space.

In general, the pure state can be shown to have the projective property [Tak54, CFZ98],

**Lemma 4.**

$$f_a \star f_b = \frac{1}{h}\, \delta_{a,b}\, f_a. \tag{25}$$

The normalization matters [Tak54]: despite linearity of the equations, it prevents naive superposition of solutions. (Quantum mechanical interference works differently here, in comportance with conventional density-matrix formalism.)

By virtue of (16), for different $\star$-genfunctions, the above dictates that

$$\int dpdx\, fg = 0. \tag{26}$$

Consequently, unless there is zero overlap for all such WFs, at least one of the two must go negative someplace to offset the positive overlap [HOS84, Coh95] — an illustration of the salutary feature of negative-valuedness. Here, this feature is *an asset and not a liability.*

Further note that integrating (17) yields the expectation of the energy,

$$\int H(x,p) f(x,p)\, dxdp = E \int f\, dxdp = E. \qquad (27)$$

N.B. Likewise, integrating the above projective condition yields

$$\int dxdp\, f^2 = \frac{1}{h}, \qquad (28)$$

which goes to a divergent result in the classical limit, for unit-normalized $f$s, as the pure state WFs grow increasingly spiky.

This discussion applies to proper WFs, (4), corresponding to *pure state* density matrices. For example, a sum of two WFs similar to a sum of two classical distributions is not a pure state in general, and so does not satisfy the condition (6). For such mixed-state generalizations, the *impurity* is [Gro46] $1 - h\langle f \rangle = \int dxdp\, (f - hf^2) \geq 0$, where the inequality is only saturated into an equality for a pure state. For instance, for $w \equiv (f_a + f_b)/2$ with $f_a \star f_b = 0$, the impurity is nonvanishing, $\int dxdp\, (w - hw^2) = 1/2$. A pure state affords a maximum of information; while the impurity is a measure of lack of information [Fan57, Tak54], characteristic of mixed states and decoherence [CSA09, Haa10] — it is the dominant term in the expansion of the quantum entropy around a pure term in the expansion of the quantum entropy around a pure state, [Bra94] providing a lower estimate for it. (The full quantum, von Neumann, entropy is $-\langle \ln \rho \rangle = -\int dxdpf \ln_\star(hf)$ [Zac07]).

**Exercise 3.** Define phase space points $\mathbf{z} \equiv (x,p)$, etc. Consider

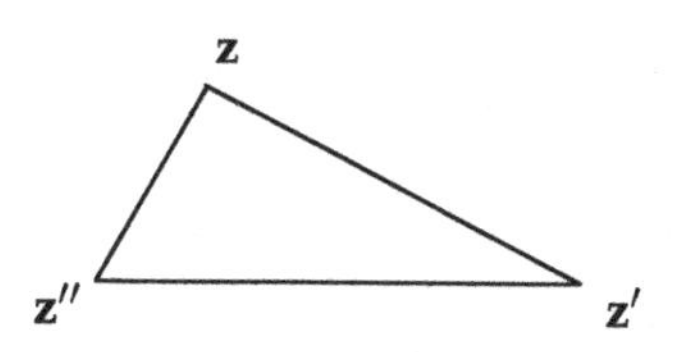


$$h(\mathbf{z}) \equiv f(\mathbf{z}) \star g(\mathbf{z}) = \int d\mathbf{z}' d\mathbf{z}'' f(\mathbf{z}') g(\mathbf{z}'')\, e^{k(\mathbf{z},\mathbf{z}',\mathbf{z}'')}.$$

What is $k(\mathbf{z},\mathbf{z}',\mathbf{z}'')$? Is it related to the area of the triangle $\Delta(\mathbf{z},\mathbf{z}',\mathbf{z}'')$? How? [Zac00].

**Exercise 4.** Prove Lagrange's representation of the shift operator, $e^{a\partial_x} f(x) = f(x+a)$, possibly using the Fourier representation, or else action on monomials $x^n$. Now, evaluate $e_\star^{ax} \star e_\star^{bp}$. Evaluate $\delta(x) \star \delta(p)$. Evaluate $e^{ax+bp} \star e^{cx+dp}$. Evaluate $(\delta(x)\,\delta(p)) \star (\delta(x)\,\delta(p))$.

**Exercise 5.** Evaluate $G(x,p) \equiv e_\star^{ax \star p}$. Hint: Show $G \star x \propto x \star G$; find the proportionality constant; solve the first-order differential equation in $\partial_p \ldots$; impose the boundary condition.

# 5. The Uncertainty Principle

In classical (non-negative) probability distribution theory, expectation values of non-negative functions are likewise non-negative, and thus yield standard *constraint inequalities* for the constituent pieces of such functions, such as, e.g. moments of the variables.

But it was just stressed that, for WFs $f$ which go negative, for an arbitrary function $g$, the expectation $\langle |g|^2 \rangle$ need not be $\geq 0$. This can be easily illustrated by choosing the support of $g$ to lie mostly in those (small) regions of phase space where the WF $f$ is negative.

Still, such constraints are not lost for WFs. It turns out they are replaced by

**Lemma 5.**

$$\langle g^* \star g \rangle \geq 0. \tag{29}$$

In Hilbert space operator formalism, this relation would correspond to the positivity of the norm. This expression is non-negative because it involves a real non-negative integrand for a pure state WF satisfying the above projective condition,[s]

$$\int dpdx(g^* \star g)f = h\int dxdp(g^* \star g)(f \star f)$$
$$= h\int dxdp(f \star g^*) \star (g \star f)$$
$$= h\int dxdp|g \star f|^2. \qquad (30)$$

To produce Heisenberg's uncertainty relation [CZ01], one now only need choose

$$g = a + bx + cp, \qquad (31)$$

for arbitrary complex coefficients $a, b, c$.

The resulting positive semidefinite quadratic form is then

$$a^*a + b^*b\langle x \star x\rangle + c^*c\langle p \star p\rangle + (a^*b + b^*a)\langle x\rangle$$
$$+(a^*c + c^*a)\langle p\rangle + c^*b\langle p \star x\rangle + b^*c\langle x \star p\rangle \geq 0, \qquad (32)$$

[s]Similarly, if $f_1$ and $f_2$ are pure state WFs, the transition probability $(|\int dx\psi_1^*(x)\psi_2(x)|^2)$ between the respective states is also non-negative [OW81], manifestly by the same argument [CZ01], providing for a non-negative phase space overlap, $\int dpdx\, f_1 f_2 = (2\pi\hbar)^2 \int dxdp\, |f_1 \star f_2|^2 \geq 0$. A mixed-state $f_1$ also has a non-negative phase space overlap integral with *all* pure states $f_2$. Conversely, it is an acceptable WF if it is normalized and has a non-negative overlap integral with all pure state WFs [HOS84] i.e. if its corresponding operator is positive semidefinite: a *bona fide* density matrix.

for any $a, b, c$. The eigenvalues of the corresponding matrix are then non-negative, and thus so must be its determinant.

Given

$$x \star x = x^2, \quad p \star p = p^2, \quad p \star x = px - i\hbar/2,$$
$$x \star p = px + i\hbar/2, \tag{33}$$

and the usual quantum fluctuations

$$(\Delta x)^2 \equiv \langle (x - \langle x \rangle)^2 \rangle, \quad (\Delta p)^2 \equiv \langle (p - \langle p \rangle)^2 \rangle, \tag{34}$$

this condition on the $3 \times 3$ matrix determinant simply amounts to

$$(\Delta x)^2 \ (\Delta p)^2 \geq \hbar^2/4 + (\langle (x - \langle x \rangle)(p - \langle p \rangle) \rangle)^2, \tag{35}$$

and hence

$$\Delta x \, \Delta p \geq \frac{\hbar}{2}. \tag{36}$$

The $\hbar$ has entered into the moments' constraint through the action of the $\star$-product [CZ01].[t]

More general choices of $g$ likewise lead to diverse expectations' inequalities in phase space; e.g., in 6-dimensional phase space, the uncertainty for $g = a + bL_x + cL_y$

---

[t]Thus, closely neighboring points in phase space evidently do not represent mutually exclusive physical contingencies, so disjoint sample space points, as required for a strict probabilistic (Kolmogorov) interpretation.

requires $l(l+1) \geq m(m+1)$, and hence $l \geq m$; and so forth [CZ01, CZ02].

W. Heisenberg

For a more extensive formal discussion of moments, cf. [NO86].

**Exercise 6.** Is the normalized phase space function [NO86]

$$g = \frac{1}{2\pi\hbar} e^{-\frac{x^2+p^2}{2\hbar}} \left( \frac{x^2+p^2}{\hbar} - 1 \right)$$

a bona fide WF? Hint: For the ground state of the oscillator, $f_0$, is $\int dx dp \; g f_0 \geq 0$? Do the second moments of $g$ satisfy the uncertainty principle?

# 6. Ehrenfest's Theorem

Moyal's equation (10),

$$\frac{\partial f}{\partial t} = \{\!\{H, f\}\!\}, \tag{37}$$

serves to prove Ehrenfest's theorem for the evolution of expectation values, often utilized in correspondence principle discussions.

For any phase space function $k(x, p)$ with no explicit time-dependence,

$$\begin{aligned} \frac{d\langle k\rangle}{dt} &= \int dx dp \, \frac{\partial f}{\partial t} k \\ &= \frac{1}{i\hbar} \int dx dp \, (H \star f - f \star H) \star k \\ &= \int dx dp \, f \{\!\{k, H\}\!\} = \langle \{\!\{k, H\}\!\} \rangle. \end{aligned} \tag{38}$$

(Any Heisenberg picture convective time-dependence, $\int dxdp\,(\dot{x}\partial_x\,(fk) + \dot{p}\,\partial_p(fk))$, would amount to an ignorable surface term, $\int dxdp(\partial_x(\dot{x}fk) + \partial_p(\dot{p}fk))$, by the $x, p$ equations of motion in that picture. Note the characteristic sign difference between the correspondent to Heisenberg's evolution equation for observables,

$$\frac{dk}{dt} = \{\!\{k, H\}\!\}, \tag{39}$$

and Moyal's equation above — in Scrödinger's picture. The $x, p$ equations of motion in such a Heisenberg picture, then, would reduce to the classical ones of Hamilton, $\dot{x} = \partial_p H$, $\dot{p} = -\partial_x H$.)

Moyal [Moy49] stressed that his eponymous quantum evolution equation (10) contrasts to Liouville's theorem (collisionless Boltzmann equation) for classical phase space densities,

$$\frac{df_{cl}}{dt} = \frac{\partial f_{cl}}{\partial t} + \dot{x}\partial_x f_{cl} + \dot{p}\partial_p f_{cl} = 0. \tag{40}$$

Specifically, unlike its classical counterpart, in general, $f$ *does not flow like an incompressible fluid in phase space*, thus depriving physical phase space trajectories of meaning, in this context.

For an arbitrary region $\Omega$ about some representative point in phase space,

**Lemma 6.**

$$\begin{aligned}\frac{d}{dt}\int_\Omega dxdp\, f &= \int_\Omega dxdp \left(\frac{\partial f}{\partial t} + \partial_x(\dot{x}f) + \partial_p(\dot{p}f)\right) \\ &= \int_\Omega dxdp(\{\!\{H, f\}\!\} - \{H, f\}) \neq 0. \end{aligned} \tag{41}$$

P. Ehrenfest

That is, the phase space region does not conserve in time the number of points swarming about the representative point: points diffuse away, in general, without maintaining the density of the quantum quasi-probability fluid; and, conversely, they are not prevented from coming together, in contrast to deterministic flow behavior. Still, *for infinite* $\Omega$ *encompassing the entire phase space,* both surface terms above vanish to yield a time-invariant normalization for the WF.

The $O(\hbar^2)$ higher momentum derivatives of the WF present in the MB (but absent in the PB — higher space derivatives probing nonlinearity in the potential) modify the Liouville flow into characteristic quantum configurations [KZZ02, FBA96, ZP94, DVS06, SKR13].

# 7. Illustration: The Harmonic Oscillator

To illustrate the formalism on a simple prototype problem, one may look at the harmonic oscillator. In the spirit of this picture, in fact, one can eschew solving the Schrödinger problem and plugging the wavefunctions into (4). Instead, for $H = (p^2 + x^2)/2$ (scaled to $m = 1$, $\omega = 1$; i.e. with $\sqrt{m\omega}$ absorbed into $x$ and into $1/p$, and $1/\omega$ into $H$), one may solve (17) directly,

$$\left(\left(x + \frac{i\hbar}{2}\partial_p\right)^2 + \left(p - \frac{i\hbar}{2}\partial_x\right)^2 - 2E\right) f(x,p) = 0. \tag{42}$$

For this Hamiltonian, then, the equation has collapsed to two simple Partial Differential Equations.

The first one, the $\Im$maginary part,

$$(x\partial_p - p\partial_x)f = 0, \tag{43}$$

restricts $f$ to depend on only one variable, the scalar in phase space,

$$z \equiv \frac{4}{\hbar}H = \frac{2}{\hbar}(x^2 + p^2). \tag{44}$$

Thus the second one, the $\Re$eal part, is a simple Ordinary Differential Equation,

$$\left(\frac{z}{4} - z\partial_z^2 - \partial_z - \frac{E}{\hbar}\right) f(z) = 0. \tag{45}$$

Setting $f(z) = \exp(-z/2)L(z)$ yields Laguerre's equation,

$$\left(z\partial_z^2 + (1-z)\partial_z + \frac{E}{\hbar} - \frac{1}{2}\right) L(z) = 0. \tag{46}$$

It is solved by Laguerre polynomials,

$$L_n = \frac{1}{n!} e^z \partial_z^n (e^{-z} z^n), \tag{47}$$

for $n = E/\hbar - 1/2 = 0, 1, 2, \ldots$, so that the $\star$-gen-Wigner-functions are [Gro46]

$$f_n = \frac{(-1)^n}{\pi\hbar} e^{-2H/\hbar} L_n\left(\frac{4H}{\hbar}\right);$$

$$L_0 = 1, \quad L_1 = 1 - \frac{4H}{\hbar},$$

$$L_2 = \frac{8H^2}{\hbar^2} - \frac{8H}{\hbar} + 1, \ldots \tag{48}$$

But for the Gaussian ground state, they all have zeros and go negative in some region.

**Lemma 7.** *Their sum provides a resolution of the identity [Moy49],*

$$\sum_{n=0}^{\infty} f_n = \frac{1}{h}. \tag{49}$$

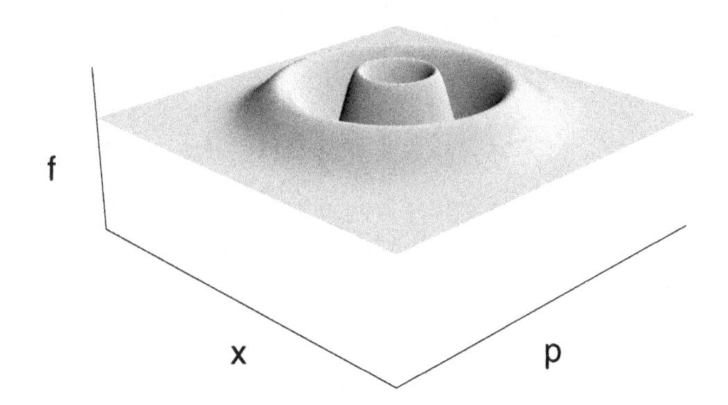


Fig. 2. The oscillator WF for the third excited state $f_3$. Note the axial symmetry, the negative values, and the nodes.

These Wigner functions, $f_n$, become spiky in the classical limit $\hbar \to 0$; e.g. the ground state Gaussian $f_0$ goes to a $\delta$-function. Since, for given $f_n$s, $\langle x^2+p^2\rangle = \hbar(2n+1)$, these become "macroscopic" for very large $n = O(\hbar^{-1})$.

Note that the energy variance, the quantum fluctuation, is

$$\langle H \star H\rangle - \langle H\rangle^2 = (\langle H^2\rangle - \langle H\rangle^2) - \frac{\hbar^2}{4}, \tag{50}$$

vanishing for all $\star$-genstates; while the naive star-less fluctuation on the right-hand side is thus larger than that, $\hbar^2/4$, and would suggest broader dispersion, groundlessly.

(For the rest of this section, scale to $\hbar = 1$, for algebraic simplicity.)

Dirac's Hamiltonian factorization method for the alternate algebraic solution of this same problem carries through intact, with $\star$-multiplication now supplanting operator multiplication. That is to say,

$$H = \frac{1}{2}(x - ip) \star (x + ip) + \frac{1}{2}. \tag{51}$$

This motivates the definition of raising and lowering functions (not operators)

$$a \equiv \frac{1}{\sqrt{2}}(x + ip), \quad a^\dagger \equiv a^* = \frac{1}{\sqrt{2}}(x - ip), \tag{52}$$

where

$$a \star a^\dagger - a^\dagger \star a = 1. \tag{53}$$

The annihilation functions $\star$-annihilate the $\star$-Fock vacuum,

$$a \star f_0 = \frac{1}{\sqrt{2}}(x + ip) \star e^{-(x^2+p^2)} = 0. \tag{54}$$

Thus, the associativity of the $\star$-product permits the customary ladder spectrum generation [CFZ98]. The $\star$-genstates for $H \star f = f \star H$ are then

$$f_n = \frac{1}{n!}(a^\dagger \star)^n f_0 (\star a)^n. \tag{55}$$

They are manifestly real, like the Gaussian ground state, and left–right symmetric. It is easy to see that they are $\star$-orthogonal for different eigenvalues. Likewise, they

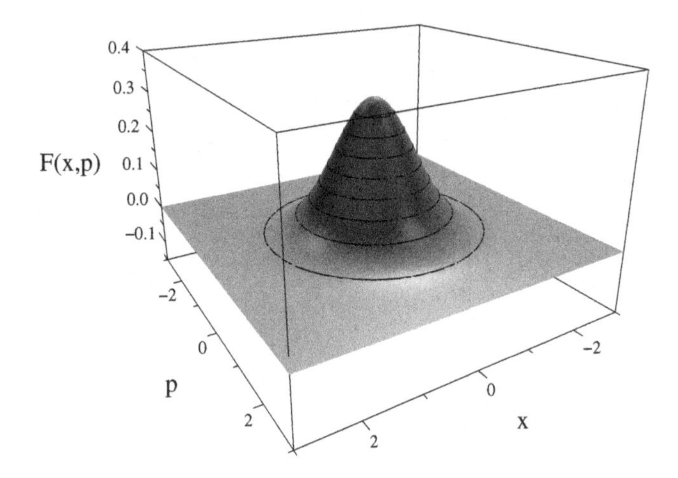


Fig. 3. The ground state $f_0$ of the harmonic oscillator, a Gaussian in phase space. It is the only $\star$-genstate with no negative values.

can be seen by the evident algebraic normal ordering to project to themselves, since the Gaussian ground state does, $f_0 \star f_0 = f_0/h$.

The corresponding coherent state WFs [HKN88, Sch88, CUZ01, Har01, DG80] are likewise analogous to the conventional formulation, amounting to this Gaussian ground state with a displacement from the phase space origin.

This type of ladder analysis carries over well to a broader class of problems [CFZ98] with "essentially isospectral" pairs of partner potentials, connected with each other through Darboux transformations relying on

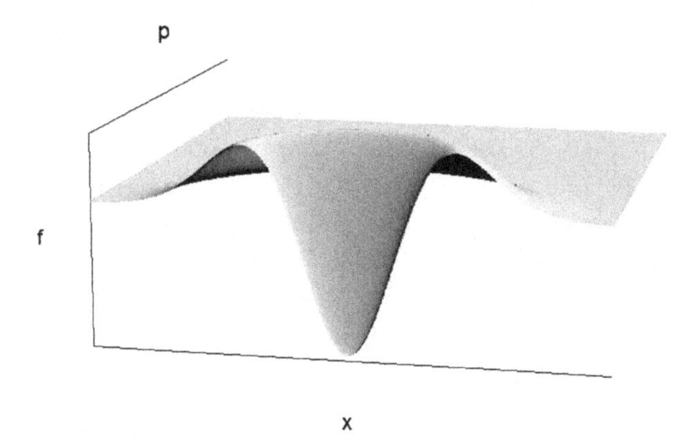


Fig. 4. Section of the oscillator WF for the first excited state $f_1$. Note the negative values. For this WF, $\langle z \rangle = 6$, where $z \equiv 2(x^2 + p^2)/\hbar$, as in the text.
On this plot, by contrast, a "classical mechanics" oscillator of energy $3\hbar/2$ would appear as a spike at a point of $z = 6$ (beyond the ridge at $z = 3$), with its phase rotating uniformly. A uniform collection (ensemble) of such rotating oscillators of all phases, or a time average of one such a classical oscillator, would be present as a stationary $\delta$-function-ring at $z = 6$.

Witten superpotentials $W$ (cf. the Pöschl–Teller potential [Ant01, APW02]). It closely parallels the standard differential operator structure of the recursive technique. That is, the pairs of related potentials and corresponding $\star$-genstate Wigner functions are constructed recursively [CFZ98] through ladder operations analogous to the algebraic method outlined above for the oscillator.

Beyond such recursive potentials, examples of further simple systems where the $\star$-genvalue equations can be solved from first principles include the linear potential [GM80, CFZ98, TZM96], the exponential interaction Liouville potentials and their supersymmetric Morse generalizations [CFZ98], and well-potential and $\delta$-function limits [KW05] (Also see [Fra00, LS82, DS82, CH86, HL99, KL94, BW10]).

Further systems may be handled through the Chebyshev-polynomial numerical techniques of [HMS98, SLC11].

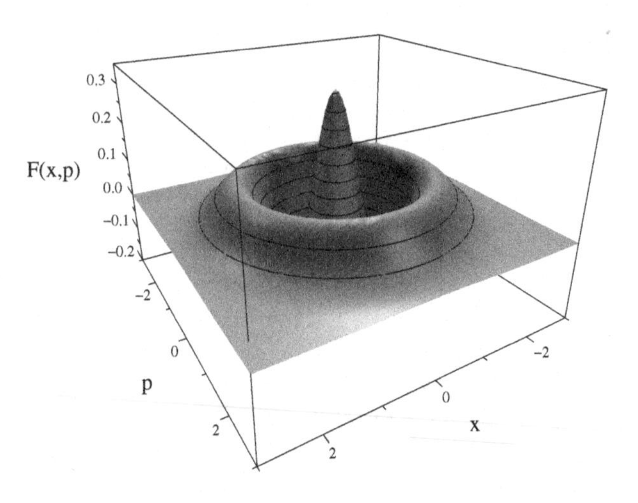


Fig. 5. The second excited state $f_2$.

A first principles phase space solution of the hydrogen atom is neither straightforward nor complete. The reader is referred to [BFF78, Bon84, DS82, CH87] for significant partial results.

Algebraic methods of generating spectra of quantum integrable models are summarized in [CZ02].

# 8. Time Evolution

Moyal's equation (10) is formally solved by virtue of associative combinatoric operations essentially analogous to Hilbert space quantum mechanics, through definition of a $\star$-unitary evolution operator, a "$\star$-exponential" [Imr67, GLS68, BFF78],

$$
\begin{aligned}
U_\star(x,p;t) &= e_\star^{itH/\hbar} \\
&\equiv 1 + (it/\hbar)H(x,p) + \frac{(it/\hbar)^2}{2!} H \star H \\
&\quad + \frac{(it/\hbar)^3}{3!} H \star H \star H + \cdots , \qquad (56)
\end{aligned}
$$

for arbitrary Hamiltonians.

The solution to Moyal's equation, given the WF at $t = 0$, then, is

**Lemma 8.**

$$f(x,p;t) = U_\star^{-1}(x,p;t) \star f(x,p;0) \star U_\star(x,p;t). \qquad (57)$$

In general, just like any $\star$-function of $H$, the $\star$-exponential (56) resolves spectrally [Bon84],

$$\exp_\star\left(\frac{it}{\hbar}H\right) = \exp_\star\left(\frac{it}{\hbar}H\right)\star 1 = \exp_\star\left(\frac{it}{\hbar}H\right)\star 2\pi\hbar\sum_n f_n$$
$$= 2\pi\hbar\sum_n e^{itE_n/\hbar} f_n, \tag{58}$$

which is thus a generating function for the $f_n$s. Of course, for $t=0$, the obvious identity resolution is recovered.

In turn, any particular $\star$-genfunction is projected out of this generating function formally by

$$\int dt \exp_\star\left(\frac{it}{\hbar}(H-E_m)\right)$$
$$= (2\pi\hbar)^2\sum_n \delta(E_n - E_m) f_n \propto f_m, \tag{59}$$

which is manifestly seen to be a $\star$-function.

**Lemma 9.** *For harmonic oscillator $\star$-genfunctions, the $\star$-exponential* (58) *is directly seen to sum to*

$$\exp_\star\left(\frac{itH}{\hbar}\right) = \left(\cos\left(\frac{t}{2}\right)\right)^{-1}\exp\left(\frac{2i}{\hbar}H\tan\left(\frac{t}{2}\right)\right), \tag{60}$$

*which is to say just a Gaussian* [BM49, Imr67, BFF78] *in phase space.*

**Corollary 1.** As a trivial application of the above, the celebrated hyperbolic tangent $\star$-composition law of Gaussians follows, since these amount to $\star$-exponentials with additive time intervals, $\exp_\star(tf) \star \exp_\star(Tf) = \exp_\star((t+T)f)$ [BFF78].

That is,

$$\exp\left(-\frac{a}{\hbar}(x^2+p^2)\right) \star \exp\left(-\frac{b}{\hbar}(x^2+p^2)\right)$$
$$= \frac{1}{1+ab}\exp\left(-\frac{a+b}{\hbar(1+ab)}(x^2+p^2)\right), \quad (61)$$

whence

$$e^{a(x^2+p^2)/\hbar} \star e^{b(x^2+p^2)/\hbar} \star e^{c(x^2+p^2)/\hbar}$$
$$= \frac{\exp\left(\frac{a+b+c+abc}{1+(ab+bc+ca)}(x^2+p^2)/\hbar\right)}{1+(ab+bc+ca)}, \quad (62)$$

and so on, with the general coefficient of $(x^2+p^2)/\hbar$ being $\tanh(\operatorname{arctanh}(a) + \operatorname{arctanh}(b) + \operatorname{arctanh}(c) + \operatorname{arctanh}(d) + \cdots)$, similar to the composition of rapidities.

N.B. This time-evolution $\star$-exponential (58) for the harmonic oscillator may be evaluated alternatively [BFF78] without explicit knowledge of the individual $\star$-genfunctions $f_n$ summed above. Instead, for (56), $U(H,t) \equiv \exp_\star(itH/\hbar)$, Laguerre's equation emerges again,

$$\partial_t U = \frac{i}{\hbar} H \star U = i\left(\frac{H}{\hbar} - \frac{\hbar}{4}(\partial_H + H\partial_H^2)\right) U, \quad (63)$$

and is readily solved by (60). One may then simply read off in the generating function (58) the $f_n$s as the Fourier-expansion coefficients of $U$.

For the variables $x$ and $p$, in the Heisenberg picture, the evolution equations collapse to mere *classical* trajectories for the oscillator,

$$\frac{dx}{dt} = \frac{x \star H - H \star x}{i\hbar} = \partial_p H = p, \tag{64}$$

$$\frac{dp}{dt} = \frac{p \star H - H \star p}{i\hbar} = -\partial_x H = -x, \tag{65}$$

where the concluding members of these two equations only hold for the oscillator, however.

Thus, for the oscillator,

$$x(t) = x\cos t + p\sin t, \quad p(t) = p\cos t - x\sin t. \tag{66}$$

As a consequence, for the harmonic oscillator, the functional form of the Wigner function is preserved along classical phase space trajectories [Gro46],

$$f(x, p; t) = f(x\cos t - p\sin t,\ p\cos t + x\sin t; 0). \tag{67}$$

*Any oscillator WF configuration rotates uniformly* on the phase plane around the origin,[u] non-dispersively: essentially classically, (cf. Fig. 6), even though it provides a

[u]This rigid rotation amounts to Condon's [Con37] continuous Fourier transform group, the Fractional Fourier Transform of signal processing [Alm94].

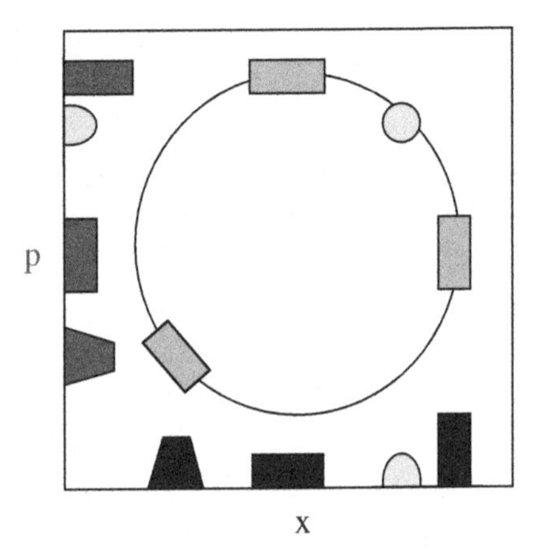


Fig. 6. Time evolution of generic WF configurations driven by an oscillator Hamiltonian. As time advances, the WF configurations rotate rigidly clockwise about the origin of phase space. (The sharp angles of the WFs in the illustration are actually unphysical, and were only chosen to monitor their "spreading wavepacket" projections more conspicuously.) These $x$- and $p$-projections (shadows) are meant to be intensity profiles on those axes, but are expanded on the plane to aid visualization. The circular figure portrays a coherent state (a Gaussian displaced off the origin) which projects on either axis identically at all times, thus without shape alteration of its wavepacket through time evolution.

complete quantum mechanical description [Gro46, BM49, Wig32, Les84, CZ99, ZC99].

Naturally, this rigid rotation in phase space preserves areas, and thus automatically illustrates the uncertainty principle. By contrast, in general, in the conventional formulation of quantum mechanics, this result is deprived of visualization import, or, at the very least, simplicity: upon integration in $x$ (or $p$) to yield the usual marginal probability densities, the rotation induces apparent complicated shape variations of the oscillating probability density profile, such as wavepacket spreading (as evident in the shadow projections on the $x$- and $p$-axes of Fig. 6), at least temporarily.

Only when (as is the case for coherent states [Sch88, CUZ01, HSD95, Sam00]) a Wigner function configuration has an *additional* axial $x-p$ symmetry around its *own* center will it possess an invariant profile upon this rotation, and hence a shape-invariant oscillating probability density [ZC99].

In Dirac's interaction representation, a more complicated interaction Hamiltonian superposed on that of the oscillator leads to shape changes of the WF configurations placed on the above "turntable", and serves to generalize to scalar field theory [CZ99].

# 9. Non-Diagonal Wigner Functions

More generally, to represent all operators on phase space in a selected basis, one looks at the Wigner-correspondents of arbitrary $|a\rangle\langle b|$, referred to as *non-diagonal WFs* [Gro46]. These enable investigation of interference phenomena and the transition amplitudes in the formulation of quantum mechanical perturbation theory [BM49, WO88, CUZ01].

Both the diagonal and the non-diagonal WFs are represented in (2), by replacing $\boldsymbol{\rho} \to |\psi_a\rangle\langle\psi_b|$,

$$\begin{aligned} f_{ba}(x,p) &\equiv \frac{1}{2\pi}\int dy\; e^{-iyp}\left\langle x+\frac{\hbar}{2}y\middle|\,\psi_a\right\rangle\,\left\langle \psi_b\middle|\,x-\frac{\hbar}{2}y\right\rangle \\ &= \frac{1}{2\pi}\int dy e^{-iyp}\psi_b^*\left(x-\frac{\hbar}{2}y\right)\psi_a\left(x+\frac{\hbar}{2}y\right) \\ &= f_{ab}^*(x,p) \\ &= \psi_a(x)\star\delta(p)\star\psi_b^*(x), \end{aligned} \tag{68}$$

(NB. The *second* index is acted upon on the left.) The representation on the last line is due to [Bra94] and lends itself to a more compact and elegant proof of Lemma 3.

Just as pure state diagonal WFs obey a projection condition, so too do the non-diagonals. For wavefunctions which are orthonormal for discrete state labels, $\int dx\, \psi_a^*(x)\psi_b(x) = \delta_{ab}$, the transition amplitude collapses to

$$\int dxdp\, f_{ab}(x,p) = \delta_{ab}. \tag{69}$$

To perform spectral operations analogous to those of Hilbert space, it is useful to note that these WFs are $\star$-orthogonal [Fai64]

$$(2\pi\hbar)\, f_{ba} \star f_{dc} = \delta_{bc} f_{da}, \tag{70}$$

as well as complete [Moy49] for integrable functions on phase space,

$$(2\pi\hbar) \sum_{a,b} f_{ab}(x_1,p_1)\, f_{ba}(x_2,p_2) = \delta(x_1 - x_2)\, \delta(p_1 - p_2). \tag{71}$$

For example, for the SHO in one dimension, non-diagonal WFs are

$$f_{kn} = \frac{1}{\sqrt{n!k!}}(a^*\star)^n\, f_0(\star a)^k,$$
$$f_0 = \frac{1}{\pi\hbar} e^{-(x^2+p^2)/\hbar}, \tag{72}$$

(cf. coherent states [CUZ01, Sch88, DG80]). The $f_{0n}$ are readily identifiable [BM49, GLS68], up to a phase space Gaussian ($f_0$), with the analytic Bargmann representation of wavefunctions. Note that

$$(a^* \star)^n f_0 = f_0 (2a^*)^n, \tag{73}$$

mere functions free of operators, where $a^* = a^\dagger$, amounts to Bargmann's variable $z$. (Further note the limit $L_0^n = 1$ below.)

Explicitly, in terms of associated Laguerre polynomials, these are [Gro46, BM49, Fai64]

$$f_{kn} = \sqrt{\frac{k!}{n!}} e^{i(k-n)\arctan(p/x)} \frac{(-1)^k}{\pi\hbar} \left( \frac{x^2+p^2}{\hbar/2} \right)^{(n-k)/2}$$
$$\times L_k^{n-k} \left( \frac{x^2+p^2}{\hbar/2} \right) e^{-(x^2+p^2)/\hbar}. \tag{74}$$

These SHO non-diagonal WFs are direct solutions to [Fai64]

$$H \star f_{kn} = E_n f_{kn}, \quad f_{kn} \star H = E_k f_{kn}. \tag{75}$$

The resulting energy $\star$-genvalue conditions are $(E_n - \frac{1}{2})/\hbar = n$, an integer, and $(E_k - \frac{1}{2})/\hbar = k$, also an integer.

The general spectral theory of WFs is covered in [BFF78, FM91, Lie90, BDW99, CUZ01].

**Exercise 7.** Consider the phase space portrayal of the simplest two-state system consisting of equal parts of oscillator ground and first excited states. Implement the above to evaluate the corresponding rotating WF: $(f_{00} + f_{11})/2 + \Re(\exp(-it)\, f_{01})$. (See Fig. 7)

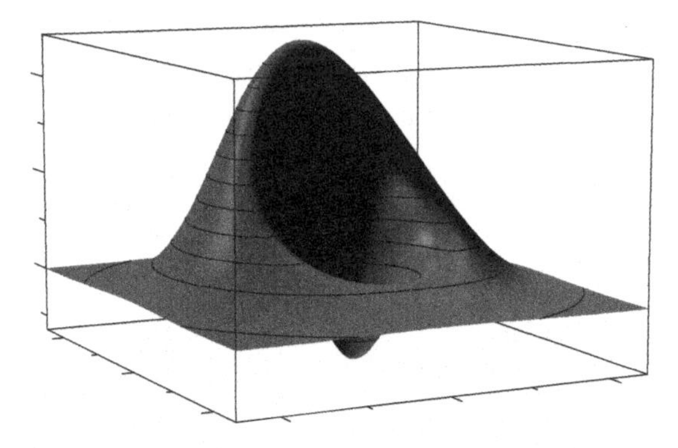

Fig. 7. Wigner Function for the superposition of the ground and first excited states of the harmonic oscillator. This simplest two-state system rotates rigidly with time.

# 10. Stationary Perturbation Theory

Given the spectral properties summarized, the phase space perturbation formalism is self-contained, and it need not make reference to the parallel Hilbert space treatment [BM49, WO88, CUZ01, SS02, MS96].

For a perturbed Hamiltonian,

$$H(x,p) = H_0(x,p) + \lambda H_1(x,p), \tag{76}$$

seek a formal series solution,

$$f_n(x,p) = \sum_{k=0}^{\infty} \lambda^k f_n^{(k)}(x,p), \quad E_n = \sum_{k=0}^{\infty} \lambda^k E_n^{(k)}, \tag{77}$$

of the left–right $\star$-genvalue equations (17), $H \star f_n = E_n f_n = f_n \star H$.

Matching powers of $\lambda$ in the eigenvalue equation [CUZ01],

$$E_n^{(0)} = \int dxdp\, f_n^{(0)}(x,p) H_0(x,p),$$

$$E_n^{(1)} = \int dxdp\, f_n^{(0)}(x,p) H_1(x,p), \tag{78}$$

$$f_n^{(1)}(x,p) = \sum_{k\neq n} \frac{f_{kn}^{(0)}(x,p)}{E_n^{(0)} - E_k^{(0)}} \times \int dXdP\, f_{nk}^{(0)}(X,P)H_1(X,P) + \sum_{k\neq n} \frac{f_{nk}^{(0)}(x,p)}{E_n^{(0)} - E_k^{(0)}} \times \int dXdP\, f_{kn}^{(0)}(X,P)H_1(X,P). \quad (79)$$

**Example.** Consider all polynomial perturbations of the harmonic oscillator in a unified treatment, by choosing

$$H_1 = e^{\gamma x+\delta p} = e_\star^{\gamma x+\delta p} = (e^{\gamma x} \star e^{\delta p})e^{i\gamma\delta/2} = (e^{\delta p} \star e^{\gamma x})e^{-i\gamma\delta/2}, \quad (80)$$

to evaluate a generating function for all the first-order corrections to the energies [CUZ01],

$$E^{(1)}(s) \equiv \sum_{n=0}^{\infty} s^n E_n^{(1)} = \int dxdp \sum_{n=0}^{\infty} s^n f_n^{(0)} H_1, \quad (81)$$

hence

$$E_n^{(1)} = \frac{1}{n!} \frac{d^n}{ds^n} E^{(1)}(s)\bigg|_{s=0}. \quad (82)$$

From the spectral resolution (58) and the explicit form of the $\star$-exponential of the oscillator Hamiltonian (60) (with $e^{it} \to s$ and $E_n^{(0)} = (n + \frac{1}{2})\hbar$), it follows that

$$\sum_{n=0}^{\infty} s^n f_n^{(0)} = \frac{1}{\pi\hbar(1+s)} \exp\left(\frac{x^2+p^2}{\hbar}\frac{s-1}{s+1}\right), \tag{83}$$

and hence

$$\begin{aligned} E^{(1)}(s) &= \frac{1}{\pi\hbar(1+s)} \int dx dp\, e^{\gamma x + \delta p} \\ &\quad \times \exp\left(-\frac{x^2+p^2}{\hbar}\frac{1-s}{1+s}\right) \\ &= \frac{1}{1-s} \exp\left(\frac{\hbar}{4}(\gamma^2+\delta^2)\frac{1+s}{1-s}\right). \end{aligned} \tag{84}$$

For example, specifically,

$$\begin{aligned} E_0^{(1)} &= \exp\left(\frac{\hbar}{4}(\gamma^2+\delta^2)\right), \\ E_1^{(1)} &= \left(1 + \frac{\hbar}{2}(\gamma^2+\delta^2)\right) E_0^{(1)}, \\ E_2^{(1)} &= \left(1 + \hbar(\gamma^2+\delta^2) + \frac{\hbar^2}{8}(\gamma^2+\delta^2)^2\right) E_0^{(1)}, \end{aligned} \tag{85}$$

and so on. All the first-order corrections to the energies are even functions of the parameters: only even functions of $x$ and $p$ can contribute to first-order shifts in the harmonic oscillator energies.

First-order corrections to the WFs may be similarly calculated using generating functions for non-diagonal WFs. Higher-order corrections are straightforward but tedious. Degenerate perturbation theory also admits an autonomous formulation in phase space, equivalent to Hilbert space and path-integral treatments.

# 11. Propagators and Canonical Transformations

Time evolution of general WFs beyond the above treatment is addressed at length in [BM49, Tak54, deB73, Ber75, GM80, CL83, OM95, CUZ01, BR93, BDR04, Wo82, Wo02, FM03, TW03, DVS06, Gat07, SKR13].

A further application of the spectral techniques outlined is the computation of the WF time-evolution operator from the propagator for wavefunctions, which is given as a bilinear sum of energy eigenfunctions,

$$\begin{aligned} G(x, X; t) &= \sum_a \psi_a(x) e^{-iE_a t/\hbar} \psi_a^*(X) \\ &\equiv \exp(iA_{\text{eff}}(x, X; t)), \end{aligned} \tag{86}$$

as it may be thought of as an exponentiated effective action. (Henceforth in this section, we scale to $\hbar = 1$.)

This leads directly to a similar bilinear double sum for the WF time-transformation kernel [Moy49],

$$\begin{aligned} &T(x, p; X, P; t) \\ &\quad = 2\pi \sum_{a,b} f_{ba}(x, p) e^{-i(E_a - E_b)t} f_{ab}(X, P). \end{aligned} \tag{87}$$

Defining a "big star" operation as a $\star$-product for the upper-case (initial) phase space variables,

$$\bigstar \equiv e^{\frac{i\hbar}{2}(\overleftarrow{\partial}_X \overrightarrow{\partial}_P - \overleftarrow{\partial}_P \overrightarrow{\partial}_X)}, \tag{88}$$

it follows that

$$\begin{aligned} &T(x,p;X,P;t)\bigstar f_{dc}(X,P) \\ &\quad = \sum_b f_{bc}(x,p)e^{-i(E_c-E_b)t} f_{db}(X,P), \end{aligned} \tag{89}$$

hence, cf. (57), propagation amounts to

$$\begin{aligned} &\int dX dP\, T(x,p;X,P;t) f_{dc}(X,P) \\ &\quad = f_{dc}(x,p)e^{-i(E_c-E_d)t} \\ &\quad = U_\star^{-1} \star f_{dc}(x,p;0) \star U_\star = f_{dc}(x,p;t). \end{aligned} \tag{90}$$

The evolution kernel $T$ thus propagates an arbitrary WF through [BM49]

$$f(x,p;t) = \int dX dP\, T(x,p;X,P;t)\, f(X,P;0). \tag{91}$$

**Exercise 8.** Utilizing the integral representation (14), $U_\star^{-1}(t) \star f(x,p;0) \star U_\star(t)$ reduces to eight integrals. Collapse four of them to obtain the above $T(x,p;X,P;t)$ as a twisted convolution of $U_\star^{-1}$ with $U_\star$ through a familiar exponential kernel. Confirm your answer with $U_\star$ for the oscillator (60), or the trivial one of the free particle, which should comport with the bottom line of the following example.

**Example.** For a free particle of unit mass in one dimension (plane wave), $H = p^2/2$, WFs propagate through the phase space kernel,

$$
\begin{aligned}
&T_{\text{free}}(x, p; X, P; t) \\
&\quad = \frac{1}{2\pi} \int dk \int dq\, e^{i(k-q)x} \delta\left(p - \frac{1}{2}(k+q)\right) \\
&\qquad \times e^{-i(q^2-k^2)t/2} e^{-i(k-q)X} \delta\left(P - \frac{1}{2}(k+q)\right) \\
&\quad = \delta(x - X - Pt)\, \delta\,(p - P),
\end{aligned}
\tag{92}
$$

identifiable as "classical" free motion,

$$
f(x, p; t) = f(x - pt, p; 0). \tag{93}
$$

The shape of any WF configuration maintains its $p$-profile, while shearing in $x$, by an amount linear in the time and $p$.

**Exercise 9.** Consider what happens to a Gaussian in phase space centered at the origin [KW90] (like the oscillator ground state $f_0$) in the absence of forces, by applying this formula. This describes the free "spreading wavepacket" of the conventional dispersive wave picture. It starts out $x - p$ symmetric, but does it stay that way? What is its asymptotic form for large times? How do you understand the "squeezing" deformation? What correlations develop in, e.g. 3-d?

**Exercise 10.** Any distribution with the special parabolic dependence $f(x,p;0) = g(x+p^2)$ will thus evolve freely as $f(x,p;t) = g((x-t^2/4)+(p-t/2)^2)$. Check that this satisfies Moyal's evolution equation (10). Since its shape merely translates rigidly in phase space, it might appear as some sort of a packet which does not spread! But, can it be normalizable? Such an unnormalizable WF of a pure state, the Airy wavetrain [CFZ98] results out of an Airy "wavefunction" which accelerates undistorted, but is not normalizable, like plane waves [BB79].

The underlying phase space structure of the evolution kernel $T(x,p;X,P;t)$ is more evident if one of the wavefunction propagators is given in coordinate space, and the other in momentum space. Then the path integral expressions for the two propagators can be combined into a single phase space path integral. For every time increment, phase space is integrated over to produce the new Wigner function from its immediate ancestor. The result is

$$
\begin{aligned}
T(x,p;X,P;t) = \frac{1}{\pi^2} \int dx_1 dp_1 \int dx_2 dp_2 e^{2i(x-x_1)(p-p_1)} \\
\times e^{-ix_1p_1} \langle x_1;t|x_2;0\rangle \langle p_1;t|p_2;0\rangle^* \\
\times e^{ix_2p_2} e^{-2i(X-x_2)(P-p_2)}, \qquad (94)
\end{aligned}
$$

where $\langle x_1;t|x_2;0\rangle$ and $\langle p_1;t|p_2;0\rangle$ are the path integral expressions in coordinate space and in momentum space.

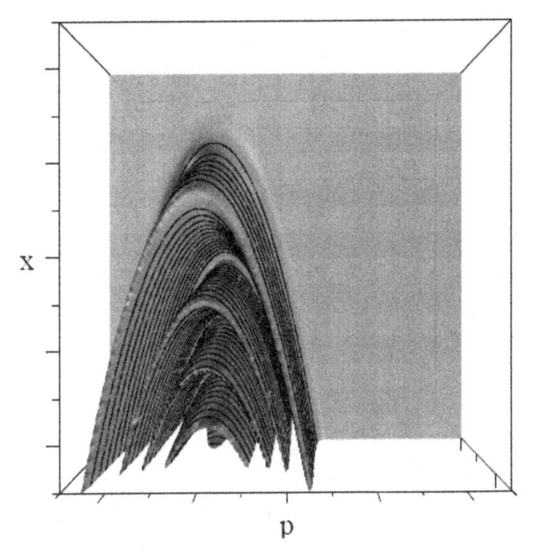


Fig. 8. The Airy wavetrain $f(x,p;t) = \frac{1}{2^{1/3}\pi}\mathrm{Ai}(2^{1/3}(x+p^2-pt))$ propagating freely while preserving its shape.

Blending these $x$ and $p$ path integrals gives a genuine path integral over phase space [Ber80, Mar91, DK85]. For a direct connection of $U_\star$ to this integral, see [Sha79, Lea68, Sam00].

Canonical transformations $(x,p) \mapsto (X(x,p), P(x,p))$ preserve the phase space volume (area) element (again, scale to $\hbar = 1$) through a trivial Jacobian,

$$dXdP = dxdp\,\{X,P\}, \tag{95}$$

i.e. they preserve Poisson Brackets

$$\{u, v\}_{xp} \equiv \frac{\partial u}{\partial x}\frac{\partial v}{\partial p} - \frac{\partial u}{\partial p}\frac{\partial v}{\partial x}, \tag{96}$$

$$\{X, P\}_{xp} = 1, \quad \{x, p\}_{XP} = 1. \tag{97}$$

Upon quantization, the c-number function Hamiltonian transforms "classically", $\mathcal{H}(X, P) \equiv H(x, p)$, like a scalar. Does the $\star$-product remain invariant under this transformation?

Yes, for *linear* canonical transformations [HKN88], but clearly *not for general canonical transformations.* Still, things can be put right, by devising general *covariant* transformation rules for the $\star$-product [CFZ98]: the WF transforms in comportance with Dirac's quantum canonical transformation theory [Dir33].

In conventional quantum mechanics, for classical canonical transformations generated by $F_{\rm cl}(x, X)$,

$$p = \frac{\partial F_{\rm cl}(x, X)}{\partial x}, \quad P = -\frac{\partial F_{\rm cl}(x, X)}{\partial X}, \tag{98}$$

the energy eigenfunctions transform in a generalization of the "representation-changing" Fourier transform [Dir33],

$$\psi_E(x) = N_E \int dX \, e^{iF(x,X)} \Psi_E(X). \tag{99}$$

(In this expression, the generating function $F$ may contain $\hbar$ corrections [BCT82] to the classical one, in general — but for several simple quantum mechanical systems it manages not to [CG92, DG02].) Hence [CFZ98], there is a transformation functional for WFs, $\mathcal{T}(x,p;X,P)$, such that

$$f(x,p) = \int dXdP\, \mathcal{T}(x,p;X,P)\bigstar\mathcal{F}(X,P)$$
$$= \int dXdP\, \mathcal{T}(x,p;X,P)\mathcal{F}(X,P), \qquad (100)$$

where

$$\mathcal{T}(x,p;X,P) = \frac{|N|^2}{2\pi}\int dYdy$$
$$\times \exp\biggl(-iyp + iPY - iF^*\left(x-\frac{y}{2}, X-\frac{Y}{2}\right)$$
$$+ iF\left(x+\frac{y}{2}, X+\frac{Y}{2}\right)\biggr). \qquad (101)$$

Moreover, it can be shown that [CFZ98],

$$H(x,p)\star\mathcal{T}(x,p;X,P) = \mathcal{T}(x,p;X,P)\bigstar\mathcal{H}(X,P). \qquad (102)$$

That is, if $\mathcal{F}$ satisfies a $\bigstar$-genvalue equation, then $f$ satisfies a $\star$-genvalue equation with the same eigenvalue, and vice versa. This proves useful in constructing WFs for simple systems which can be trivialized classically through canonical transformations.

A thorough discussion of MB automorphisms may start from [BCW02]. (Also see [Hie82, DKM88, GR94, OM95, DV97, Hak99, KL99, DP01].)

Dynamical time evolution is also a canonical transformation [Dir33], with the generator's role played by the effective action $A_{\rm eff}$ introduced above, incorporating quantum corrections to both phases and normalizations. It propagates initial wavefunctions to those at a final time.

**Example.** For the linear potential with $m = 1/2$,

$$H = p^2 + x, \tag{103}$$

wavefunction evolution is determined by the propagator

$$\exp(iA_{\rm lin}(x, X; t)) = \frac{1}{\sqrt{4\pi i t}} \exp\left(\frac{i(x-X)^2}{4t} - \frac{i(x+X)t}{2} - \frac{it^3}{12}\right). \tag{104}$$

$T$ then evaluates to

$$\begin{aligned}
T_{\rm lin}(x, p; X, P; t) &= \frac{1}{2\pi} \int dY dy \\
&\quad \times \exp\left(-iyp + iPY - iA^*_{\rm lin}\left(x - \frac{y}{2}, X - \frac{Y}{2}; t\right)\right. \\
&\quad \left. + iA_{\rm lin}\left(x + \frac{y}{2}, X + \frac{Y}{2}; t\right)\right) \\
&= \frac{1}{8\pi^2 t} \int dY dy \exp\left(-iyp + iPY - \frac{it}{2}(y+Y)\right. \\
&\quad \left. + \frac{i}{2t}(x-X)(y-Y)\right)
\end{aligned}$$

$$= \frac{1}{2t}\delta\left(p + \frac{t}{2} - \frac{x-X}{2t}\right)\delta\left(P - \frac{t}{2} - \frac{x-X}{2t}\right)$$

$$= \delta(p + t - P)\,\delta\,(x - 2tp - t^2 - X)$$

$$= \delta(x - X - (p + P)t)\,\delta\,(P - p - t). \tag{105}$$

The $\delta$-functions enforce exactly the classical motion for a mass $= 1/2$ particle subject to a negative constant force of unit magnitude (acceleration $= -2$). Thus the WF evolves "classically" as

$$f(x, p; t) = f(x - 2pt - t^2, p + t; 0). \tag{106}$$

NB. Time-*in*dependence follows for $f(x, p; 0)$ being any function of the energy variable, since it stays constant, $x + p^2 = x - 2pt - t^2 + (p + t)^2$.

# 12. The Weyl Correspondence

This section summarizes the formal bridge and equivalence of phase space quantization to the conventional operator formulation of quantum mechanics in Hilbert space. The Weyl correspondence merely provides a change of representation between phase space and Hilbert space. In itself, it does not map (commutative) classical mechanics to (non-commutative) quantum mechanics ("quantization"), as Weyl had originally hoped. But it makes the deformation map at the heart of quantization easier to grasp, now defined within a common representation, and thus more intuitive.

Weyl [Wey27] introduced an association rule mapping, invertibly, c-number phase space functions $g(x,p)$ (called phase space kernels) to operators $\mathfrak{G}$ in a given ordering prescription. Specifically, $p \mapsto \mathfrak{p}$, $x \mapsto \mathfrak{x}$, and, in general,

$$\mathfrak{G}(\mathfrak{x},\mathfrak{p}) = \frac{1}{(2\pi)^2}\int d\tau d\sigma dx dp\, g(x,p) \times \exp(i\tau(\mathfrak{p}-p)+i\sigma(\mathfrak{x}-x)). \qquad (107)$$

The eponymous ordering prescription requires that an arbitrary operator, regarded as a power series in $\mathfrak{x}$ and $\mathfrak{p}$, be first-ordered in a completely symmetrized expression in $\mathfrak{x}$ and $\mathfrak{p}$, by use of Heisenberg's commutation relations, $[\mathfrak{x}, \mathfrak{p}] = i\hbar$.

A term with $m$ powers of $\mathfrak{p}$ and $n$ powers of $\mathfrak{x}$ is obtained from the coefficient of $\tau^m\sigma^n$ in the expansion of $(\tau\mathfrak{p} + \sigma\mathfrak{x})^{m+n}$, which serves as a generating function of Weyl-ordered polynomials [GF91]. It is evident how the map yields a Weyl-ordered operator from a polynomial phase space kernel. It includes every possible ordering with multiplicity one, e.g.

$$6p^2x^2 \longmapsto \mathfrak{p}^2\mathfrak{x}^2 + \mathfrak{x}^2\mathfrak{p}^2 + \mathfrak{p}\mathfrak{x}\mathfrak{p}\mathfrak{x} + \mathfrak{p}\mathfrak{x}^2\mathfrak{p} + \mathfrak{x}\mathfrak{p}\mathfrak{x}\mathfrak{p} + \mathfrak{x}\mathfrak{p}^2\mathfrak{x}. \qquad (108)$$

In general [McC32],

$$p^m x^n \longmapsto \frac{1}{2^n}\sum_{r=0}^{n}\binom{n}{r}\mathfrak{x}^r\mathfrak{p}^m\mathfrak{x}^{n-r}$$
$$= \frac{1}{2^m}\sum_{s=0}^{m}\binom{m}{s}\mathfrak{p}^s\mathfrak{x}^n\mathfrak{p}^{m-s}. \qquad (109)$$

Phase space constants map to the constant multiplying $\mathbb{1}$, the identity in Hilbert space.

**Exercise 11.** Weyl-order $x^3p^2$, i.e. find its Weyl map. How many terms are there? can you find an equivalent re-expression with fewer terms, and no explicit $\hbar$'s, using Heisenberg's commutation relation?

In this correspondence scheme, then,

$$h\,\mathrm{Tr}\mathfrak{G} = \int dxdp\; g. \qquad (110)$$

Conversely [Dir30, Gro46, Kub64, Lea68, HOS84], the c-number phase space kernels $g(x,p)$ of Weyl-ordered operators $\mathfrak{G}(\mathfrak{x},\mathfrak{p})$ are specified by $\mathfrak{p} \mapsto p$, $\mathfrak{x} \mapsto x$; or, more precisely, by the "Wigner map",

$$g(x,p) = \frac{\hbar}{2\pi} \int d\tau d\sigma e^{i(\tau p+\sigma x)} \operatorname{Tr}(e^{-i(\tau\mathfrak{p}+\sigma\mathfrak{x})}\mathfrak{G})$$
$$= \hbar \int dy\, e^{-iyp} \left\langle x + \frac{\hbar}{2}y \middle| \mathfrak{G}(\mathfrak{x},\mathfrak{p}) \middle| x - \frac{\hbar}{2}y \right\rangle, \tag{111}$$

since the above trace, in the coordinate representation, $\exp(i\tau\mathfrak{p})|x\rangle = |x - \hbar\tau\rangle$, reduces to

$$\int dz e^{i\tau\sigma\hbar/2} \langle z|e^{-i\sigma\mathfrak{x}} e^{-i\tau\mathfrak{p}}\mathfrak{G}|z\rangle$$
$$= \int dz e^{i\sigma(\tau\hbar/2-z)} \langle z - \hbar\tau|\mathfrak{G}|z\rangle. \tag{112}$$

Equivalently, the c-number integral kernel of the operator amounts to [Dir30, Bas48],

**Lemma 10.**

$$\langle x|\mathfrak{G}|y\rangle = \int \frac{dp}{2\pi\hbar} \exp\left(ip\frac{(x-y)}{\hbar}\right) g\left(\frac{x+y}{2}, p\right). \tag{113}$$

**Exercise 12.** For the SHO, note the standard evolution amplitude $\langle x|\exp(-it\mathfrak{H}/\hbar)|0\rangle$, so the propagator $G(x,0;t)$,

(86), follows by just inserting (60)* for $g$ into, and evaluating this integral.

Thus, the density matrix $|\psi_b\rangle\langle\psi_a|/h$ inserted in this expression [Moy49] yields the hermitian generalization of the Wigner function (68) encountered,

$$\begin{aligned}
f_{ab}(x,p) &\equiv \frac{1}{2\pi}\int dy\, e^{-iyp}\left\langle x+\frac{\hbar}{2}y\middle|\,\psi_b\right\rangle\left\langle\psi_a\middle|\, x-\frac{\hbar}{2}y\right\rangle \\
&= \frac{1}{2\pi}\int dy e^{-iyp}\psi_a^*\left(x-\frac{\hbar}{2}y\right)\psi_b\left(x+\frac{\hbar}{2}y\right) \\
&= \frac{1}{(2\pi)^2}\int d\tau d\sigma\,\langle\psi_a|\, e^{i\tau(\mathfrak{p}-p)+i\sigma(\mathfrak{x}-x)}\,|\psi_b\rangle \\
&= f_{ba}^*(x,p), \qquad (114)
\end{aligned}$$

where the $\psi_a(x)$s are (ortho-) normalized solutions of a Schrödinger problem. (Wigner [Wig32] mainly considered the diagonal elements of the pure state density matrix, denoted above as $f_m \equiv f_{mm}$.)

As a consequence, matrix elements of operators, i.e. traces of them with the density matrix, are obtained through mere phase space integrals [Moy49, Bas48],

$$\langle\psi_m|\mathfrak{G}|\psi_n\rangle = \int dx dp\, g(x,p) f_{mn}(x,p), \qquad (115)$$

and thus expectation values follow for $m=n$, as utilized throughout in this overview.

Hence, above all,

**Lemma 11.**

$$\langle\psi_m|\exp i(\sigma\mathfrak{x}+\tau\mathfrak{p})|\psi_m\rangle$$
$$=\int dxdp\ f_m(x,p)\exp i(\sigma x+\tau p), \qquad (116)$$

the celebrated *moment-generating functional* [Moy49, Bas48] of the Wigner distribution, codifying the expectation values of *all* moments.

Products of Weyl-ordered operators are not necessarily Weyl-ordered, but may be easily reordered into unique Weyl-ordered operators through the degenerate Campbell–Baker–Hausdorff identity.[v] In a study of the uniqueness of the Schrödinger representation, von Neumann [Neu31] adumbrated the composition rule of kernel functions in such operator products, appreciating that Weyl's correspondence was in fact a homomorphism. (Effectively, he arrived at the Fourier-space convolution representation of the star product below; equivalently, at the detailed parameterization of the Heisenberg group representation involved.)

Finally, Groenewold [Gro46] neatly worked out in detail how the kernel functions (i.e. the Wigner transforms) $f$ and $g$ of two operators $\mathfrak{F}$ and $\mathfrak{G}$ must compose to yield the

[v]This amounts to the specification of Weyl's representation of the Heisenberg group.

kernel (the Wigner map image, sometimes called the "Weyl symbol") of the product $\mathfrak{F}\mathfrak{G}$,

$$\mathfrak{F}\mathfrak{G} = \frac{1}{(2\pi)^4}\int d\xi d\eta d\xi' d\eta' dx' dx'' dp' dp'' f(x',p') g(x'',p'')$$
$$\times \exp i(\xi(\mathfrak{p}-p') + \eta(\mathfrak{x}-x'))$$
$$\times \exp i(\xi'(\mathfrak{p}-p'') + \eta'(\mathfrak{x}-x''))$$
$$= \frac{1}{(2\pi)^4}\int d\xi d\eta d\xi' d\eta' dx' dx'' dp' dp'' f(x',p') g(x'',p'')$$
$$\times \exp i((\xi+\xi')\mathfrak{p} + (\eta+\eta')\mathfrak{x})$$
$$\times \exp i\left(-\xi p' - \eta x' - \xi' p'' - \eta' x'' + \frac{\hbar}{2}(\xi\eta' - \eta\xi')\right). \tag{117}$$

Changing integration variables to

$$\xi' \equiv \frac{2}{\hbar}(x-x'), \quad \xi \equiv \tau - \frac{2}{\hbar}(x-x'),$$
$$\eta' \equiv \frac{2}{\hbar}(p'-p), \quad \eta \equiv \sigma - \frac{2}{\hbar}(p'-p), \tag{118}$$

reduces the above integral to the *fundamental isomorphism*,

**Theorem**

$$\mathfrak{F}\mathfrak{G} = \frac{1}{(2\pi)^2}\int d\tau d\sigma dx dp$$
$$\times \exp i(\tau(\mathfrak{p}-p) + \sigma(\mathfrak{x}-x))(f\star g)(x,p), \tag{119}$$

where $f\star g$ is the expression *(13)*.

Noncommutative operator multiplication Wigner-transforms to $\star$-multiplication.

The $\star$-product thus specifies the transition from classical to quantum mechanics.

In fact, the failure of Weyl-ordered operators to close under multiplication may be stood on its head [Bra03], to *define* a Weyl-symmetrizing operator product, which is commutative and associative and constitutes the Weyl transform of $fg$ instead of the noncommutative $f \star g$. (For example,

$$2x \star p = 2xp + i\hbar \mapsto 2\mathfrak{x}\mathfrak{p} = \mathfrak{x}\mathfrak{p} + \mathfrak{p}\mathfrak{x} + i\hbar. \tag{120}$$

The classical piece of $2x \star p$ maps to the Weyl-symmetrization of the operator product, $2xp \mapsto \mathfrak{x}\mathfrak{p} + \mathfrak{p}\mathfrak{x}$.) One may then solve for the PB in terms of the MB, and, through the Weyl correspondence, reformulate Classical Mechanics in Hilbert space as a deformation of Quantum Mechanics, instead of the other way round [Bra03]!

Arbitrary operators $\mathfrak{G}(\mathfrak{x}, \mathfrak{p})$ consisting of operators $\mathfrak{x}$ and $\mathfrak{p}$ in various orderings but with the same classical limit, could be *imagined* to be rearranged by use of Heisenberg commutations to canonical completely symmetrized Weyl-ordered forms, in general with $O(\hbar)$ terms generated in the process.

Trivially, each one might then be inverse-transformed uniquely to its Weyl-correspondent c-number kernel function $g$ in phase space. (However, in practice [Kub64], there is the above more direct Wigner transform formula (111), which bypasses any need for an actual explicit rearrangement. Since operator products amount to convolutions of such matrix-element integral kernels, $\langle x|\mathfrak{G}|y\rangle$, explicit reordering issues can be systematically avoided.)

Thus, operators differing from each other by different orderings of their $\mathfrak{x}$s and $\mathfrak{p}$s Wigner-map to kernel functions $g$ coinciding with each other at $O(\hbar^0)$ but not at $O(\hbar)$, in general. Hence, in phase space quantization, a survey of all alternate operator orderings in a problem with such ambiguities amounts to a survey of the "quantum correction" $O(\hbar)$ pieces of the respective kernel functions, i.e. the Wigner transforms of those operators, and their accounting is often systematized and expedited.

Choice-of-ordering problems then reduce to purely $\star$-product algebraic ones, as the resulting preferred orderings are specified through particular deformations in the c-number kernel expressions resulting from the particular solution in phase space [CZ02].

**Exercise 13.** Evaluate the $\star$-genvalues $\lambda$ of $\Pi(x,p) \equiv \frac{h}{2}\delta(x)\delta(p)$. (One might think that spiky functions like this have no place in phase space quantization, but they do: one may check that this is but the phase space kernel, i.e. the Wigner transform, of the parity operator [Roy77], $\int dx|-x\rangle\langle x| = \frac{h}{2(2\pi)^2}\int d\tau d\sigma \exp(i\tau\mathfrak{p}+i\sigma\mathfrak{x})$. So, then, what is $\Pi \star \Pi$?) Hint for $\Pi \star f = \lambda f$: for the SHO basis (48), what is $\Pi \star f_0(x,p)$? And what is $\Pi \star f_1(x,p)$? What must then be their value at the origin, $x = 0 = p$? How does one then see the necessity of the overall alternating signs in that basis?

H. Weyl

# 13. Alternate Rules of Association

The Weyl correspondence rule (107) is not unique: there are a host of alternate *equivalent* association rules which specify corresponding representations. All these representations with equivalent formalisms are typified by characteristic quasi-distribution functions and $\star$-products, all systematically inter-convertible among themselves. They have been surveyed comparatively and organized in [Lee95, BJ84], on the basis of seminal classification work by Cohen [Coh66, Coh76]. Like different coordinate transformations, they may be favored by virtue of their different characteristic properties in varying applications.

For example, instead of the symmetric operator $\exp(i\tau\mathfrak{p} + i\sigma\mathfrak{x})$ underlying the Weyl transform, one might posit, instead [Lee95, HOS84], antistandard ordering,

$$\exp(i\tau\mathfrak{p})\exp(i\sigma\mathfrak{x}) = \exp(i\tau\mathfrak{p} + i\sigma\mathfrak{x})\ w(\tau, \sigma), \qquad (121)$$

with $w = \exp(i\hbar\tau\sigma/2)$, which specifies the Kirkwood–Rihaczek prescription [Kir33]; or else standard ordering (momenta to the right), $w = \exp(-i\hbar\tau\sigma/2)$ instead on the right-hand-side of the above, for the "Mehta" prescription, also utilized by Moyal [Moy49, Blo40, Yv46]; or their (real) average, $w = \cos(\hbar\tau\sigma/2)$ for the older

Rivier prescription [Ter37]; or normal and antinormal orderings for the Glauber–Sudarshan prescriptions, generalizing to $w = \exp(\frac{\hbar}{4}(\tau^2 + \sigma^2))$ for the Husimi prescription [Hus40, Tak89, Ber80] which is underlain by coherent states; or $w = \sin(\hbar\tau\sigma/2)/(\hbar\tau\sigma/2)$, for the Born–Jordan prescription; and so on.

**Exercise 14.** The standard ordering prescription [Ter37, Blo40] was used early on for its simplicity, $f_M(x,p) = \psi^*(x)\phi(p)\exp(ipx/\hbar)/\sqrt{2\pi\hbar}$, where $\phi(p) \equiv \int dx \exp(-ixp/\hbar)\,\psi(x)/\sqrt{2\pi\hbar}$. Show [Moy49, Yv46] that the Wigner function is readily obtainable from it, $f(x,p) = e^{-i\hbar\partial_x\partial_p/2}\, f_M(x,p)$.

The corresponding quasi-distribution functions in each representation can be obtained systematically as convolution transforms of each other [Coh76, Lee95, HOS84]; and, likewise, the kernel function observables are convolution "dressings" of each other, as are their $\star$-products [Dun88, AW70, Ber75, Ber80].

**Example.** For instance, the (normalized) Husimi distribution follows from a "Gaussian smoothing" (Gaussian low-pass filtering, or Weierstrass transform) invertible linear conversion map [Ber80, WO87, Tak89, Lee95, AMP09] of the WF,

$$f_H = T(f) = \exp\left(\frac{\hbar}{4}(\partial_x^2 + \partial_p^2)\right) f$$

$$= \frac{1}{\pi\hbar} \int dx' dp'$$

$$\times \exp\left(-\frac{(x'-x)^2 + (p'-p)^2)}{\hbar}\right) f(x', p'), \qquad (122)$$

and likewise for the observables. (So, for instance, the oscillator hamiltonian now becomes $H_H = (p^2 + x^2 + \hbar)/2$, slightly nonclassical.) Thus, for the *same operators* $\mathfrak{G}$, in this alternate ordering,

$$\langle \mathfrak{G} \rangle = \int dx dp\, g(x,p) \exp\left(-\frac{\hbar}{4}(\partial_x^2 + \partial_p^2)\right) f_H$$

$$= \int dx dp\ g_H\ e^{\hbar(\overleftarrow{\partial}_x \overrightarrow{\partial}_x + \overleftarrow{\partial}_p \overrightarrow{\partial}_p)/2} f_H. \qquad (123)$$

That is, expectation values of observables now entail equivalence conversion dressings of the respective kernel functions — and a corresponding isomorph ⋆-product [Ba79, OW81, Vor89, Tak89, Zac00],

$$\circledast = \exp\left(\frac{\hbar}{2}(\overleftarrow{\partial}_x \overrightarrow{\partial}_x + \overleftarrow{\partial}_p \overrightarrow{\partial}_p)\right) \star$$

$$= \exp\left(\frac{\hbar}{2}(\overleftarrow{\partial}_x - i \overleftarrow{\partial}_p)(\overrightarrow{\partial}_x + i \overrightarrow{\partial}_p)\right), \qquad (124)$$

cf. (131) below.

Evidently, however, this $\circledast$ now *cannot be simply dropped inside integrals*, quite *unlike* the case of the WF (16).

For this reason, quantum distributions such as this Husimi distribution (which is actually [deB67, Car76, OW81,Jan84,Ste80] positive semidefinite[w] — and in a very restricted class of distributions with that property [Bas86]) *cannot* be automatically thought of as *bona fide* distribution functions, in some contrast to the WF — which is thus a bit of a "first among equals" in this respect [Yv46].

This is often dramatized as the failure of the Husimi distribution $f_H$ to yield the correct $x$- or $p$-marginal probabilities, upon integration by $p$ or $x$, respectively [OW81, HOS84]. Since phase space integrals are thus complicated by conversion dressing convolutions, they preclude direct implementation of the Schwarzin equality and the standard inequality-based moment-constraining techniques of probability theory, as well as routine completeness- and orthonormality-based functional-analytic operations.

Ignoring the above equivalence dressings and, instead, simply treating the Husimi distribution as an ordinary probability distribution in evaluating expectation values,

[w]This is evident from the factorization of the constituent integrals of $f_H(0,0)$ to a complex norm squared, or, more directly, the first footnote of Sec. 5 since the Gaussian is $f_0$ for the harmonic oscillator; and hence at *all points* in phase space.

nevertheless results in loss of quantum information — effectively "coarse-graining" (low-pass filtering) to a semiclassical limit, and thereby increasing the relevant entropy [Bra94].

**Exercise 15.** In this Husimi representation, show that $f_H$ is normalized to 1. For its oscillator $H_H$, show that $H_H \circledast f_{H\,n} = E_{H\,n} f_{H\,n}$. Is this differential equation in $z$ *simpler* than in the Wigner representation? (What order in $z$ is it?) Hence, find the simple (un-normalized) $f_H$'s. Alternatively, solve for $U_H$ in $\hbar\partial_t U_H = iH_H \circledast U_H$, and thence read off these simple $f_H$'s.

Similar caveats also apply to more recent symplectic tomographic representations [MMT96, MMM01, Leo97], which are also positive semidefinite, but also do not quite constitute conventional probability distributions.

**Exercise 16.** One may work out Moyal's inter-relations [Moy49, Yv46, Coh66, Coh76] between the Weyl-ordering kernel (Wigner transform) functions and the standard-ordering correspondents, as well as the respective dressing relations between the proper $\star$-products [Lee95], in systematic analogy to the foregoing example for the Husimi prescription. The weight $w = \exp(-i\hbar\tau\sigma/2)$ mentioned dictates a dressing of kernels, $g_s = T(g) \equiv \exp(-i\hbar\partial_x\partial_p/2)\, g(x,p)$, and of $\star$-products by (131) below.

Further abstracting the Weyl-map functional of Sec. 12, for generic Hilbert space variables $\mathfrak{z}$ and phase space variables $z$, the Weyl map compacts to an integral kernel [Kub64], $\mathfrak{G}(\mathfrak{z}) = \int dz \Delta(\mathfrak{z}, z) g(z)$, and the inverse (Wigner) map to $g(z) = h\text{Tr}(\overline{\Delta}(\mathfrak{z}, z)\mathfrak{G}(\mathfrak{z}))$. Here, $h\text{Tr}(\Delta(\mathfrak{z}, z)\overline{\Delta}(\mathfrak{z}, z')) = \delta^2(z - z')$, $\int dz \Delta(\mathfrak{z}, z) = \int dz \overline{\Delta}(\mathfrak{z}, z) = \mathbb{1}$, and $h\text{Tr}\Delta = h\text{Tr}\overline{\Delta} = 1$.

The $\star$-product is thus a convolution in the integral representation, cf. (13),

**Lemma 12.**

$$f \star g = \int dz' dz'' f(z') g(z'') \ h\text{Tr}(\overline{\Delta}(\mathfrak{z}, z)\Delta(\mathfrak{z}, z')\Delta(\mathfrak{z}, z'')). \tag{125}$$

The dressing of these functionals is then given by $\Delta_s(\mathfrak{z}, z) = T^{-1}(z)\Delta(\mathfrak{z}, z)$, so that both prescriptions yield the *same operator* $\mathfrak{G}$, when $g_s(z) = T(z)g(z)$ and $\overline{\Delta}_s = T\overline{\Delta}$.

Thus, more abstractly, the corresponding integral kernel for $\circledast$ amounts to just $h\text{Tr}(T(z)\overline{\Delta}(\mathfrak{z}, z)T^{-1}(z')\Delta(\mathfrak{z}, z')$ $T^{-1}(z'')\Delta(\mathfrak{z}, z''))$.

# 14. The Groenewold–van Hove Theorem; The Uniqueness of MBs and $\star$-Products

Groenewold's correspondence principle theorem [Gro46] (to which van Hove's extension to all association rules is often attached [vH51, AB65, Ar83]) enunciates that, in general, there is *no invertible linear map* from *all functions* of phase space $f(x,p), g(x,p), \ldots$, to hermitian operators in Hilbert space $\mathfrak{Q}(f)$, $\mathfrak{Q}(g), \ldots$, such that the PB structure is preserved,

$$\mathfrak{Q}(\{f,g\}) = \frac{1}{i\hbar}[\mathfrak{Q}(f), \mathfrak{Q}(g)], \tag{126}$$

as envisioned in Dirac's ("functor") heuristics [Dir25].

Instead, the Weyl correspondence map (107) from functions to ordered operators,

$$\mathfrak{W}(f) \equiv \frac{1}{(2\pi)^2} \int d\tau d\sigma dx dp f(x,p) \\ \times \exp(i\tau(\mathfrak{p} - p) + i\sigma(\mathfrak{x} - x)), \tag{127}$$

determines the $\star$-product in (119), $\mathfrak{W}(f \star g) = \mathfrak{W}(f)\, \mathfrak{W}(g)$, and thus the Moyal Bracket Lie algebra,

$$\mathfrak{W}(\{\!\{f,g\}\!\}) = \frac{1}{i\hbar}[\mathfrak{W}(f), \mathfrak{W}(g)]. \tag{128}$$

*It is the MB, then, instead of the PB, which maps invertibly to the quantum commutator.*

That is to say, the "deformation" involved in phase space quantization is nontrivial: the quantum (observable) functions, in general, need not coincide with the classical ones [Gro46], and often involve $O(\hbar)$ corrections, as extensively illustrated, e.g. in [CZ02, DS02, CH86]; also see [Got99, Tod12].

For example, as was already discussed, the Wigner transform of the square of the angular momentum $\mathfrak{L} \cdot \mathfrak{L}$ turns out to be $L^2 - 3\hbar^2/2$, significantly for the ground-state Bohr orbit [She59, DS82, DS02].

**Lemma 13.** *Groenewold's early celebrated* counterexample *noted that the classically vanishing PB expression*

$$\{x^3, p^3\} + \frac{1}{12}\{\{p^2, x^3\}, \{x^2, p^3\}\} = 0 \qquad (129)$$

*is* anomalous *in implementing Dirac's heuristic proposal to substitute commutators of $\mathfrak{Q}(x), \mathfrak{Q}(p), \ldots,$ for PBs upon quantization: indeed, this substitution, or the equivalent substitution of MBs for PBs, yields a* Groenewold anomaly, $-3\hbar^2$, *for this specific expression.*

**Exercise 17.** Beyond Hilbert space, in phase space, check that the standard linear operator realization $\mathfrak{V}(f) \equiv i\hbar(\partial_x f \partial_p - \partial_p f \partial_x)$ satisfies (126). But is it invertible? N.B. $\mathfrak{V}(\{x, p\}) = 0$.

An alternate abstract operator realization of the above MB Lie algebra in phase space (as opposed to the Hilbert space one, $\mathfrak{W}(f)$) linearly is [FFZ89, CFZm98]

$$\mathfrak{K}(f) = f \star . \tag{130}$$

Realized on a toroidal phase space, upon a formal identification $\hbar \mapsto 2\pi/N$, this realization of the MB Lie algebra leads to the Lie algebra of $SU(N)$ [FFZ89], by means of Sylvester's clock-and-shift matrices [Syl82]. For generic $\hbar$, it may be thought of as a generalization of $SU(N)$ for continuous $N$. This allows for taking the limit $N \to \infty$, to thus contract to the PB algebra.

Essentially (up to isomorphism), the MB algebra is the unique (Lie) one-parameter deformation (expansion) of the Poisson Bracket algebra [Vey75, BFF78, FLS76, Ar83, Fle90, deW83, BCG97, TD97], a uniqueness extending to the (associative) star product.

Isomorphism allows for dressing transformations of the variables (kernel functions and WFs, as in Sec. 13 on alternate orderings), through linear maps $f \mapsto T(f)$, which leads to cohomologically equivalent star-product variants, i.e. [Ba79, Vor89, BFF78],

$$T(f \star g) = T(f) \circledast T(g). \tag{131}$$

The $\star$-MB algebra is isomorphic to the algebra of $\circledast$-MB.

Computational features of $\star$-products are addressed in [BFF78, Han84, RO92, Zac00, EGV89, Vo78, An97, Bra94].

# 15. Advanced Topic: Quasi-Hermitian Quantum Systems

So far, the discussion has limited itself to hermitian operators and systems.

However, superficially non-hermitian Hamiltonian quantum systems are also of considerable current interest, especially in the context of PT symmetric models [Ben07,Mos05], although many of the main ideas appeared earlier [SGH92,XA96]. For such systems, the Hilbert space structure is at first sight very different from that for hermitian Hamiltonian systems, inasmuch as the dual wavefunctions are *not* just the complex conjugates of the wavefunctions, or, equivalently, the Hilbert space metric is *not* the usual one. While it is possible to keep most of the compact Dirac notation in analyzing such systems, here we work with explicit functions and avoid abstract notation, in the hope to fully expose all the structure, rather than to hide it.

Many theories are "*quasi-hermitian*", as given by the entwining relation

$$\mathfrak{G}\mathfrak{H} = \mathfrak{H}^\dagger \mathfrak{G}, \tag{132}$$

where "the metric" $\mathfrak{G}$ is a hermitian, invertible, and *positive-definite* operator. All adjoints here are specified in a *pre-defined* Hilbert space, with a given scalar product and norm. Existence of such a $\mathfrak{G}$ is a necessary and sufficient condition for a completely diagonalizable $\mathfrak{H}$ to have real eigenvalues. In such situations, it is *not* necessary that $\mathfrak{H} = \mathfrak{H}^\dagger$ to yield real-energy eigenvalues.

Given $\mathfrak{H}$, there are two widely-used methods to find all such $\mathfrak{G}$:

(I) Solve the entwining relation directly (e.g. as a PDE in phase space); or,

(II) Solve for the eigenfunctions of $\mathfrak{H}$, find their biorthonormal dual functions, and then construct $\mathfrak{G} \sim(\text{dual})^\dagger \times (\text{dual})$, or $\mathfrak{G}^{-1} \sim(\text{state}) \times (\text{state})^\dagger$. In principle, these methods are equivalent. In practice, one or the other may be easier to implement.

Once a $\mathfrak{G}$ is available, an equivalent hermitian Hamiltonian is

$$\mathbb{H} = \sqrt{\mathfrak{G}}\mathfrak{H}\sqrt{\mathfrak{G}^{-1}} = \mathbb{H}^\dagger. \tag{133}$$

So, why consider apparently non-hermitian structures at all? *A priori*, one may not know that $\mathfrak{G}$ exists, let alone what it actually is. But even when one does have $\mathfrak{G}$, and finally $\mathbb{H}$, the manifestly hermitian form of an interesting model may be *non-local*, and more difficult to analyze than an equivalent, local, quasi-hermitian form of the model.

Here, we illustrate the general theory of quasi-hermitian systems in quantum phase space, for the "imaginary Liouville theory" [CV07]:

$$\mathfrak{H}(\mathfrak{x},\mathfrak{p}) = \mathfrak{p}^2 + \exp(2i\mathfrak{x}), \quad \mathfrak{H}^\dagger(\mathfrak{x},\mathfrak{p}) = \mathfrak{p}^2 + \exp(-2i\mathfrak{x}). \quad (134)$$

Several other notable applications of QMPS methods to PT symmetric models have been made [SG05, SG06, dMF06]. We scale to $\hbar = 1$.

### Solutions of the metric equation

The above entwining relation $\mathfrak{G}\mathfrak{H} = \mathfrak{H}^\dagger\mathfrak{G}$, or alternatively $\mathfrak{H}\mathfrak{G}^{-1} = \mathfrak{G}^{-1}\mathfrak{H}^\dagger$, can be written as a PDE through the use of deformation quantization techniques in phase space.

If the Weyl kernel of $\mathfrak{G}^{-1}$ is denoted by "the dual metric" $\widetilde{G}(x,p)$,

$$\mathfrak{G}^{-1}(\mathfrak{x},\mathfrak{p}) = \frac{1}{(2\pi)^2}\int d\tau d\sigma dx dp\, \widetilde{G}(x,p) \times \exp(i\tau(\mathfrak{p}-p) + i\sigma(\mathfrak{x}-x)), \quad (135)$$

then the entwining equation in phase space is

$$H(x,p) \star \widetilde{G}(x,p) = \widetilde{G}(x,p) \star \overline{H(x,p)}. \quad (136)$$

For the imaginary Liouville example, $H \star \widetilde{G} = \widetilde{G} \star \overline{H}$ boils down to the linear differential-difference equation

$$p\frac{\partial}{\partial x}\widetilde{G}(x,p) = \sin(2x)\widetilde{G}(x,p-1). \quad (137)$$

Hermitian $\mathfrak{G}^{-1}$ is represented here by a *real* Weyl kernel $\widetilde{G}$.

Basic solutions to the $H \star \widetilde{G}$ entwining relation are obtained by separation of variables. We find two classes of solutions, labeled by a parameter $s$. The *first class* of solutions is non-singular for all real $p$, although there are zeroes for negative integer $p$,

$$\widetilde{G}(x,p;s) = \frac{1}{s^p \Gamma(1+p)} \exp\left(-\frac{s}{2}\cos 2x\right). \qquad (138)$$

For real $s$, this is real and positive definite on the positive momentum half-line.

Solutions in the *other class* have poles and corresponding changes in sign for positive $p$,

$$\widetilde{G}_{\text{other}}(x,p;s) = \frac{\Gamma(-p)}{s^p} \exp\left(\frac{s}{2}\cos 2x\right). \qquad (139)$$

Linear combinations of these are also solutions of the linear entwining equation. This linearity permits us to build a particular *composite metric* from members of the first class by using a contour integral representation. Namely,

$$\widetilde{G}(x,p) \equiv \frac{1}{2\pi i} \int_{-\infty}^{(0+)} \widetilde{G}(x,p;s) \frac{e^{s/2}}{s}\, ds. \qquad (140)$$

The contour begins at $-\infty$, with $\arg s = -\pi$, proceeds below the real $s$-axis towards the origin, loops in the positive counterclockwise sense around the origin (hence the $(0+)$ notation), and then continues above the real $s$-axis back to $-\infty$, with $\arg s = +\pi$.

Evaluation of the contour integral yields

$$\widetilde{G}(x,p) = \frac{(\sin^2 x)^p}{(\Gamma(p+1))^2}, \tag{141}$$

where use is made of Sonine's contour representation of the $\Gamma$ function,

$$\frac{1}{\Gamma(1+p)} = \frac{1}{2\pi i} \int_{-\infty}^{(0+)} \tau^{-p-1} e^{\tau}\, d\tau. \tag{142}$$

**The $\star$ root of the metric**

We now look for an equivalence between the Liouville, $H = p^2 + e^{2ix}$, and the free particle, $\mathbb{H} = p^2$, as given by solutions of the entwining equation,

$$H(x,p) \star \widetilde{S}(x,p) = \widetilde{S}(x,p) \star p^2. \tag{143}$$

For the Liouville $\longleftrightarrow$ free-particle case, this amounts to a first-order PDE similar to that for $\widetilde{G}$, but inherently complex:

$$2ip \frac{\partial}{\partial x} \widetilde{S}(x,p) = e^{2ix} \widetilde{S}(x,p-1). \tag{144}$$

Once again, solutions are easily found through the use of a product *ansatz.* For any value of a parameter $s$, we also find two classes of solutions:

$$\widetilde{S}(x,p;s) = \frac{1}{s^p \Gamma(1+p)} \exp\left(-\frac{s}{4}\exp(2ix)\right),$$

$$\widetilde{S}_{\text{other}}(x,p;s) = \frac{\Gamma(-p)}{s^p} \exp\left(\frac{s}{4}\exp(2ix)\right). \tag{145}$$

The first of these is a "good" solution for $p \in (-1, \infty)$, say, while the second is good for $p \in (-\infty, 0)$, thereby providing a pair of solutions that cover the entire real $p$ axis — but *not* so easily joined together.

**The dual metric as an absolute $\star$ square**

Each such solution for $\widetilde{S}$ leads to a candidate real metric, given by

$$\widetilde{G} = \widetilde{S} \star \overline{\widetilde{S}}. \tag{146}$$

To verify this, we note that the entwining equation for $\widetilde{S}$, and its conjugate $\overline{\widetilde{S}}$,

$$H \star \widetilde{S} = \widetilde{S} \star p^2, \quad p^2 \star \overline{\widetilde{S}} = \overline{\widetilde{S}} \star \overline{H}, \tag{147}$$

may be combined with the associativity of the star product to obtain

$$H \star \widetilde{S} \star \overline{\widetilde{S}} = \widetilde{S} \star p^2 \star \overline{\widetilde{S}} = \widetilde{S} \star \overline{\widetilde{S}} \star \overline{H}. \tag{148}$$

For the first class of $\widetilde{S}$ solutions, by choosing $s = \pm 2$, and again using the standard integral representation for $1/\Gamma$, we find a result that coincides with the above composite dual metric (141),

$$\widetilde{S}(x, p; \pm 2) \star \overline{\widetilde{S}}(x, p; \pm 2) = \frac{(\sin^2 x)^p}{(\Gamma(p+1))^2} = \widetilde{G}(x, p). \tag{149}$$

This proves that the corresponding operator is positive (perhaps positive definite) and provides a greater appreciation of the $\star$ roots of $\widetilde{G}$.

## Wavefunctions and Wigner transforms

The eigenvalue problem is well-posed if wavefunctions are required to be bounded (free particle BCs) solutions to

$$\left(-\frac{\partial^2}{\partial x^2} + m^2 e^{2ix}\right) \psi_E = E\psi_E. \tag{150}$$

The coupling parameter $m$ has not been set to $m = 1$ yet, even though the free limit is not discussed.

All real $E \geq 0$ are allowed, and the solutions are doubly degenerate for $E > 0$ and $\sqrt{E}$ non-integer. This follows from making a change of variable,

$$z = me^{ix}, \tag{151}$$

to obtain Bessel's equation, and hence,

$$J_{\pm\sqrt{E}}(me^{ix}) = \left(\frac{m}{2}e^{ix}\right)^{\pm\sqrt{E}} \sum_{n=0}^{\infty} \frac{(-m^2/4)^n}{n!\,\Gamma(1+n\pm\sqrt{E})} e^{2inx}. \tag{152}$$

Note the ground state $E=0$ solution is non-degenerate, and given by $J_0(me^{ix})$. In fact, all integer $\sqrt{E}$ are also non-degenerate, since $J_{-n}(z) = (-1)^n J_n(z)$.

## Integral representations for $E=n^2$; quantum equivalence to a free particle on a circle

The $2\pi$-periodic Bessel functions are, in fact, the canonical integral transforms of free plane waves on a circle, as constructed in this special situation just by exponentiating the classical generating function. Explicitly,

$$J_n(me^{ix}) = \frac{1}{2\pi}\int_0^{2\pi} \exp(-in\theta) \times \exp(ime^{ix}\sin\theta)d\theta, \quad n \in \mathbb{Z}, \tag{153}$$

with $J_{-n}(z) = (-1)^n J_n(z)$.

The integral transform is a two-to-one map from the space of all free particle plane waves to Bessel functions: $e^{\mp in\theta} \to (\pm 1)^n J_n$. But, acting on the linear combinations $e^{in\theta} + (-1)^n e^{-in\theta}$, the kernel gives a map which is one-to-one, hence invertible on this subspace. The situation here is exactly like the real Liouville QM, for all positive energies, except for the fact that here we have a *well-behaved ground state.*

### Dual wavefunctions

The "PT method" of constructing the dual space by simply changing normalizations and phases of the wavefunctions does *not* provide a biorthonormalizable set of functions in this case, since

$$\frac{1}{2\pi}\int_0^{2\pi} J_k(me^{ix})J_n(me^{ix})dx = \begin{cases} 1 & \text{if } k = n = 0 \\ 0 & \text{otherwise} \end{cases}. \tag{154}$$

This follows because the $J$s only contain positive powers of $e^{ix}$. So, all the $2\pi$-periodic energy eigenfunctions are *self-orthogonal* except for the ground state. In retrospect, this difficulty was circumvented by Carl Neumann in the mid-19th century.

### A simple $2\pi$-periodic biorthogonal system

Elements of the dual space for the $2\pi$-periodic eigenfunctions are given by Neumann polynomials, $\{A_n\}$. For all analytic Bessel functions of non-negative integer index,

$$J_n(z) = \left(\frac{z}{2}\right)^n \sum_{k=0}^{\infty} \frac{(-1)^k}{k!(k+n)!} \left(\frac{z}{2}\right)^{2k}, \tag{155}$$

there are corresponding associated *Neumann polynomials* in powers of $1/z$ that are dual to $\{J_n\}$ on any contour enclosing the origin.

These are given by

$$A_0(z) = 1, \quad A_1(z) = \frac{2}{z},$$

$$A_{n\geq 2}(z) = n \left(\frac{2}{z}\right)^n \sum_{k=0}^{\lfloor n/2 \rfloor} \frac{(n-k-1)!}{k!} \left(\frac{z}{2}\right)^{2k}. \tag{156}$$

These $A_n$ satisfy an *inhomogeneous* equation where the inhomogeneity is orthogonal to all the $J_k(z)$:

$$-\frac{d^2}{dx^2} A_n(me^{ix}) + (m^2 e^{2ix} - n^2) A_n(me^{ix}) = \begin{cases} 2nme^{ix} & \text{for odd } n \\ 2m^2 e^{2ix} & \text{for even } n \neq 0 \end{cases}, \tag{157}$$

$$-\frac{d^2}{dx^2} J_n(me^{ix}) + (m^2 e^{2ix} - n^2) J_n(me^{ix}) = 0. \qquad (158)$$

Re-expressed for the imaginary Liouville problem, the key orthogonality reads

$$\frac{1}{2\pi} \int_0^{2\pi} A_k \left(me^{ix}\right) J_n \left(me^{ix}\right) dx = \delta_{kn}. \qquad (159)$$

Hence, as detailed below, the integral kernel of the (dual) metric, $\langle x|\mathfrak{G}^{-1}|y\rangle$, on the space of dual wavefunctions is

$$\begin{aligned} J(x,y) &\equiv J_0(me^{-ix} - me^{iy}) \\ &= \sum_{n=0}^{\infty} \varepsilon_n J_n(me^{-ix}) J_n(me^{iy}), \end{aligned} \qquad (160)$$

where $\varepsilon_0 = 1$, $\varepsilon_{n\neq 0} = 2$.

This manifestly hermitian, bilocal kernel $J(x,y) = J(y,x)^*$ can be used to evaluate the norm of a general function in the span of the eigenfunctions,

$$\psi(x) \equiv \sum_{n=0}^{\infty} c_n \sqrt{\varepsilon_n} J_n(me^{ix}), \qquad (161)$$

through use of the corresponding dual wavefunction

$$\psi_{\text{dual}}(x) \equiv \sum_{n=0}^{\infty} c_n^* A_n(me^{ix}) / \sqrt{\varepsilon_n}, \qquad (162)$$

where, once again, $\varepsilon_0 = 1$, $\varepsilon_{n\neq 0} = 2$.

The result is, as expected,

$$\|\psi\|^2 = \frac{1}{(2\pi)^2}\int_0^{2\pi} dx \int_0^{2\pi} dy\ \overline{\psi_{\text{dual}}(x)} J(x,y)\psi_{\text{dual}}(y)$$
$$= \sum_{n=0}^{\infty} |c_n|^2. \tag{163}$$

**Wigner transform of a generic bilocal metric**

In general, a scalar product for any generic biorthogonal system such as $\{A_k, J_n\}$ can be written as a double integral over configuration space involving a generic metric bilocal kernel, $\mathcal{J}(x,y)$,

$$(\phi,\psi) = \iint \phi(x)\mathcal{J}(x,y)\psi(y)dxdy. \tag{164}$$

When a scalar product is so expressed, it may be readily re-expressed in phase space through the use of a Wigner transform,

$$f_{\psi\phi}(x,p) \equiv \frac{1}{2\pi}\int e^{iyp}\psi\left(x-\frac{1}{2}y\right)\phi\left(x+\frac{1}{2}y\right)dy. \tag{165}$$

Fourier inverting gives the point-split product,

$$\phi(x)\psi(y) = \int_{-\infty}^{\infty} e^{i(y-x)p}\ f_{\psi\phi}\left(\frac{x+y}{2},p\right)dp. \tag{166}$$

Thus, the scalar product can be re-written as

$$(\phi,\psi) = \iint \mathcal{G}(x,p)f_{\psi\phi}(x,p)dxdp, \tag{167}$$

where the generic phase space metric is the Wigner transform (111) of the bilocal metric,

$$\mathcal{G}(x,p) = \int e^{iyp} \mathcal{J}\left(x - \frac{1}{2}y, x + \frac{1}{2}y\right) dy, \tag{168}$$

and inversely, (113),

$$\mathcal{J}(x,y) = \frac{1}{2\pi} \int_{-\infty}^{\infty} e^{i(x-y)p} \mathcal{G}\left(\frac{x+y}{2}, p\right) dp. \tag{169}$$

### Example: Liouville dual metric

Now, to be specific, for $2\pi$-periodic *dual* functions of imaginary Liouville quantum mechanics, the scalar product specified previously through (160) can be re-expressed for $m = 1$ in a form which is immediately convertible to phase space, through

$$J(x,y) = J_0\left(-2ie^{i(y-x)/2} \sin\left(\frac{x+y}{2}\right)\right). \tag{170}$$

The corresponding dual metric in the phase space peculiar to this example is given by the Wigner transform of this bilocal, namely,

$$\begin{aligned} \widetilde{G}(x,p) &= \frac{1}{2\pi} \int_0^{2\pi} J(x+w, x-w) e^{2iwp} dw \\ &= \frac{1}{2\pi} \int_0^{2\pi} J_0(-2ie^{-iw} \sin x) e^{2iwp} dw. \end{aligned} \tag{171}$$

Hence the simple final answer,

$$\widetilde{G}(x,p) = \frac{\left(\sin^2 x\right)^p}{(p!)^2} \tag{172}$$

for integer $p \geq 0$, but vanishes for integer $p < 0$. This is, yet again, the above solution (141) of the entwining equation.

An equivalent operator expression can be obtained by the method of Weyl transforms, (113).

# 16. Omitted Miscellany

Phase space quantization extends in several interesting directions which are not covered in such a brief introduction.

Symmetry effects of collections of identical particles are systematically described in [SchN59, Imr67, BC62, Jan78, OW84, HOS84, CBJ07]. Finite-temperature profiles embodying these quantum statistics in phase space are illustrated in [Kir33, vZy12].

Disentanglement in heat baths, the quantum Langevin equation, and quantum Brownian motion (summarized in [FO11]) are worked out in detail in [FO01, FO05, FO07, FO10].

Dynamical scattering and tunneling of wavepacket WFs off wells [Raz96, BDR04], barriers [KKFR89], Gaussian barrier potentials [SLC11] abound, especially in the numerical literature.

The systematic generalization of the $\star$-product to arbitrary non-flat Poisson manifolds [Kon97] is a culmination of extensions to general symplectic and Kähler geometries [Fed94, Mor86, CGR90, Kis01], and varied symplectic contexts [Ber75, Rie89, Bor96, KL92, RT00, Xu98, Kar98, CPP02, BGL01].

For further work on curved spaces, see [APW02, BF81, PT99]. For extensive reviews of mathematical issues, see [And69, Hor79, Fol89, Unt79, Bou99, Wo98, AW70]. For a connection to the theory of modular forms, see [Raj02].

For WFs on discrete phase spaces (finite-state systems), see, among others, in [Woo87, KP94, OBB95, ACW98, RA99, RG00, BHP02, MPS02].

Spin is treated in [Str57, deG74, Kut72, BGR91, VG89, AW00], and forays into a relativistic formulation in [LSU02] (also see [CS75, Ran66]).

Inclusion of electromagnetic fields and gauge invariance is treated in [Kub64, Mue99, BGR91, LF94, LF01, JVS87, ZC99, KO00, MP04]. Subtleties of Berry's phase in phase space are addressed in [Sam00].

Applications of the phase space quantum picture include efficient computation of $\zeta$-function regularization determinants [KT07].

# 17. Synopses of Selected Papers

The decisive contributors to the development of the formulation are Hermann Weyl (1885–1955), Eugene Wigner (1902–1995), Hilbrand Groenewold (1910–1996), and Jose Moyal (1910–1998). The bulk of the theory is implicit in Groenewold's and Moyal's seminal papers.

But confidence in the autonomy of the formulation accreted slowly and fitfully. As a result, an appraisal of critical milestones cannot avoid subjectivity. Nevertheless, here we provide summaries of a few papers that we believe remedied confusion about the logical structure of the formulation.

H. Weyl (1927) [Wey27] introduces the correspondence of "Weyl-ordered" operators to phase space (c-number) kernel functions. The correspondence is based on Weyl's formulation of the Heisenberg group, appreciated through a discrete QM application of Sylvester's (1883) [Syl82] clock and shift matrices. The correspondence is proposed as a general quantization prescription, unsuccessfully, since it fails, e.g. with angular momentum squared.

J. von Neumann (1931) [Neu31], expatiates on a Fourier transform version of the $\star$-product, in a technical aside off an analysis of the uniqueness of Schrödinger's representation, based on Weyl's Heisenberg group formulation. This then effectively promotes Weyl's correspondence rule to full isomorphism between Weyl-ordered operator multiplication and $\star$-convolution of kernel functions. Nevertheless, this result is not properly appreciated in von Neumann's celebrated own book on the Foundations of QM.

E. Wigner (1932) [Wig32], the author's first paper in English, introduces the eponymous phase space distribution function controlling quantum mechanical diffusive flow in phase space. It notes the negative values, and specifies the time evolution of this function and applies it to quantum statistical mechanics. (Actually, Dirac (1930) [Dir30] has already considered a formally identical construct, and an implicit Weyl correspondence, for the approximate electron density in a multi-electron Thomas–Fermi atom; but, interpreting negative values as a failure of that semiclassical approximation, he crucially hesitates about the full quantum object.)

H. Groenewold (1946) [Gro46], a seminal but inadequately appreciated paper, is based on Groenewold's thesis work. It achieves full understanding of the Weyl correspondence as an invertible transform, rather than as a consistent quantization rule. It articulates and recognizes the WF as the phase space (Weyl) kernel of the density matrix. It reinvents and streamlines von Neumann's construct into the standard $\star$-product, in a systematic exploration of the isomorphism between Weyl-ordered operator products and their kernel function compositions. It thus demonstrates how Poisson Brackets contrast crucially to quantum commutators — "Groenewold's Theorem". By way of illustration, it further works out the harmonic oscillator WF.

J. Moyal (1949) [Moy49] enunciates a grand synthesis: It establishes an independent formulation of quantum mechanics in phase space. It systematically studies all expectation values of Weyl-ordered operators, and identifies the Fourier transform of their moment-generating function (their characteristic function) with the Wigner Function. It further interprets the subtlety of the "negative probability" formalism and reconciles it with the uncertainty principle and the diffusion of the probability fluid. Not least, it recasts the time evolution of the Wigner Function through a deformation of the Poisson Bracket into the Moyal Bracket (the commutator of $\star$-products, i.e. the Wigner transform of the Heisenberg commutator), and thus opens up the way for a systematic study of the semiclassical limit. Before publication, Dirac contrasts this work favorably to his own ideas on functional integration, in Bohr's Festschrift [Dir45], despite private reservations and lengthy arguments with Moyal. Various subsequent scattered observations of French investigators on the statistical approach [Yv46], as well as Moyal's, are collected in J. Bass (1948) [Bas48], which further stretches to hydrodynamics. Earlier Soviet efforts include [Ter37, Blo40].

M. Bartlett and J. Moyal (1949) [BM49] applies this language to calculate propagators and transition probabilities for oscillators perturbed by time-dependent potentials.

T. Takabayasi (1954) [Tak54] investigates the fundamental projective normalization condition for pure state Wigner functions, and exploits Groenewold's link to the conventional density matrix formulation. It further illuminates the diffusion of wavepackets.

G. Baker (1958) [Bak58] (Baker's thesis paper) envisions the logical autonomy of the formulation, sustained by the projective normalization condition as a basic postulate. It resolves measurement subtleties in the correspondence principle and appreciates the significance of the anticommutator of the $\star$-product as well, thus shifting emphasis to the $\star$-product itself, over and above its commutator.

D. Fairlie (1964) [Fai64] (also see [Kun67, Coh76, Dah83, Bas48]) explores the time-independent counterpart to Moyal's evolution equation, which involves the $\star$-product, beyond mere Moyal Bracket equations, and derives (instead of postulating) the projective orthonormality conditions for the resulting Wigner functions. These now allow for a unique and full solution of the quantum system, in principle (without any reference to the conventional Hilbert space formulation). Autonomy of the formulation is fully recognized.

R. Kubo (1964) [Kub64] elegantly reviews, in modern notation, the representation change between Hilbert space and phase space — although in ostensible ignorance of Weyl's and Groenewold's specific papers. It applies the phase space picture to the description of electrons in a uniform magnetic field, initiating gauge-invariant formulations and pioneering "noncommutative geometry" applications to diamagnetism and the Hall effect.

N. Cartwright (1976) [Car76] notes that the WF smoothed by a phase space Gaussian (i.e. Weierstrass transformed) as wide as or wider than the minimum uncertainty packet is positive semidefinite. Actually, this convolution result goes further back to at least de Bruijn (1967) [deB67] and Iagolnitzer (1969) [Iag69], if not Husimi (1940) [Hus40].

M. Berry (1977) [Ber77] elucidates the subtleties of the semiclassical limit, ergodicity, integrability, and the singularity structure of Wigner function evolution. Complementary results are featured in Voros (1976–1978) [Vo78].

F. Bayen, M. Flato, C. Fronsdal, A. Lichnerowicz, and D. Sternheimer (1978) [BFF78] analyzes systematically the deformation structure and the uniqueness of the formulation, with special emphasis on spectral theory, and consolidates it mathematically. (Also see Berezin [Ber75].) It provides explicit illustrative solutions to standard problems and utilizes influential technical tools, such as the $\star$-exponential (already known in [Imr67, GLS68]).

A. Royer (1977) [Roy77] interprets WFs as the expectation value of the operators effecting reflections in phase space. (Further see [Kub64, Gro76, BV94].)

G. García-Calderón and M. Moshinsky (1980) [GM80] implements the transition from Hilbert space to phase space to extend classical propagators and canonical transformations to quantum ones in phase space. (The most conclusive work to date is [BCW02]. Further see [HKN88, Hie82, DKM88, CFZ98, DV97, GR94, Hak99, KL99, DP01].)

J. Dahl and M. Springborg (1982) [DS82] initiates a thorough treatment of the hydrogen atom and other simple atoms in phase space, albeit not from first principles — the WFs are evaluated in terms of Schrödinger wave-functions.

M. de Wilde and P. Lecomte (1983) [deW83] consolidates the deformation theory of $\star$-products and MBs on general real symplectic manifolds, analyzes their cohomology structure, and confirms the absence of obstructions.

M. Hillery, R. O'Connell, M. Scully, and E. Wigner (1984) [HOS84] has served the physics community as the classic introduction to phase-space quantization and the Wigner function.

Y. Kim and E. Wigner (1990) [KW90] is a classic pedagogical discussion of the spread of wavepackets in phase space, uncertainty-preserving transformations, and coherent and squeezed states.

B. Fedosov (1994) [Fed94] initiates an influential geometrical construction of the $\star$-product on all symplectic manifolds.

T. Curtright, D. Fairlie, and C. Zachos (1998) [CFZ98] illustrates more directly the equivalence of the time-independent $\star$-genvalue problem to the Hilbert space formulation, and hence its logical autonomy; formulates Darboux isospectral systems in phase space; works out the covariant transformation rule for general nonlinear canonical transformations (with reliance on the classic work of P. Dirac (1933) [Dir33]); and thus furnishes explicit solutions of practical problems on first principles, without recourse to the Hilbert space formulation. Efficient techniques for perturbation theory are based on generating functions for complete sets of Wigner functions in T. Curtright, T. Uematsu, and C. Zachos (2001) [CUZ01]. A self-contained derivation of the uncertainty principle in phase space is given in T. Curtright and C. Zachos (2001) [CZ01].

M. Hug, C. Menke, and W. Schleich (1998) [HMS98] introduce and exemplify techniques for numerical solution of $\star$-equations on a basis of Chebyshev polynomials. Dynamical scattering of wavepacket WFs off Gaussian barrier potentials on a similar basis is detailed in [SLC11].

# Bibliography

## References


AW70. G. Agarwal and E. Wolf, *Phys Rev* **D2** (1970) 2161; *ibid* 2187, *ibid* 2206

Alm94. L. Almeida, *IEEE Trans Sig Proc* **42** (1994) 3084–3091

APW02. M. Alonso, G. Pogosyan, and K-B. Wolf, *J Math Phys* **43** (2002) 5857 [quant-ph/0205041]

And69. R. Anderson, *J Funct Anal* **4** (1969) 240–247; *ibid* **9** (1972) 423–440

AW00. J-P. Amiet and S. Weigert, *Phys Rev* **A63** (2000) 012102

Ant01. J-P. Antoine, J-P. Gazeau, P. Monceau, J. Klauder, and K. Penson, *J Math Phys* **42** (2001) 2349 [math-ph/0012044]

An97. F. Antonsen, [gr-qc/9710021]

Ara95. T. Arai, *J Math Phys* **36** (1995) 622–630

AB65. R. Arens and D. Babbitt, *J Math Phys* **6** (1965) 1071–1075

Ar83. W. Arveson, *Commun Math Phys* **89** (1983) 77–102

ACW98. N. Atakishiyev, S. Chumakov, and K. B. Wolf, *J Math Phys* **39** (1998) 6247–6261

AMP09. A. Athanassoulis, N. Mauser, and T. Paul, *J Math Pure Appl* **91** (2009) 296–338

Bak58. G. Baker, *Phys Rev* **109** (1958) 2198–2206

Bak60. G. Baker, I. McCarthy, and C. Porter, *Phys Rev* **120** (1960) 254–264

BJ84. N. Balazs and B. Jennings, *Phys Rep* **104** (1984) 347–391

BV90. N. Balazs and A. Voros, *Ann Phys (NY)* **199** (1990) 123–140

Bal75. R. Balescu, *Equilibrium and Nonequilibrium Statistical Mechanics* (Wiley-Interscience, New York, 1975)

BRWK99. K. Banaszek, C. Radzewicz, K. Wódkiewicz, and J. Krasinski, *Phys Rev* **A60** (1999) 674–677

BBL80. H. Bartelt, K. Brenner, and A. Lohmann, *Opt Commun* **32** (1980) 32–38

Bar45. M. Bartlett, *Proc Camb Phil Soc* **41** (1945) 71–73

Bas48. J. Bass, *Rev Scientifique* **86** (3299) (1948) 643–652; *Comput Rend Acad Sci* **221** (1945) 46–49

Bas86. S. Basu, *Phys Lett* **A114** (1986) 303–305

BM49. M. Bartlett and J. Moyal, *Proc Camb Phil Soc* **45** (1949) 545–553

BKM03. I. Bars, I. Kishimoto, and Y. Matsuo, *Phys Rev* **D67** (2003) 126007

BGL01. I. Batalin, M. Grigoriev, and S. Lyakhovich, *Theor Math Phys* **128** (2001) 1109–1139 [hep-th/0101089]

BFF78. F. Bayen, M. Flato, C. Fronsdal, A. Lichnerowicz, and D. Sternheimer, *Ann Phys (NY)* **111** (1978) 61–110; *ibid* 111–151; *Lett Math Phys* **1** (1977) 521–530

BF81. F. Bayen and C. Fronsdal, *J Math Phys* **22** (1981) 1345–1349

Ba79. F. Bayen, in *Group Theoretical Methods in Physics*, W. Beiglböck *et al.* (eds.), Lecture Notes in Physics, vol. 94 (Springer-Verlag, Heidelberg, 1979), pp. 260–271

BW10. B. Belchev and M. Walton, *J Phys* **A43** (2010) 225206

BJY04. A. Belitsky, X. Ji, and F. Yuan, *Phys Rev* **D69** (2004) 074014;
A. Belitsky and A. Radyushkin, *Phys Rep* **418** (2005) 1–387

BDR04. M. Belloni, M. Doncheski, and R. Robinett, *Am J Phys* **72** (2004) 1183–1192

Ben07. C. Bender, *Rep Prog Phys* **70** (2007) 947 [hep-th/0703096]; *J Phys* **A45**(44) (2012) 9 November, Special issue on *Quantum Physics with Non-Hermitian Operators*

BC99. M. Benedict and A. Czirják, *Phys Rev* **A60** (1999) 4034–4044

BC09. G. Benenti and G. Casati, *Phys Rev* **E79** (2009) 025201

Ber80. F. Berezin, *Sov Phys Usp* **23** (1980) 763–787

Ber75. F. Berezin, *Commun Math Phys* **40** (1975) 153–174; *Math USSR Izv* **8** (1974) 1109–1165

Ber77. M. Berry, *Phil Trans Roy Soc Lond* **A287** (1977) 237–271;
M. Berry and N. Balazs, *J Phys* **A12** (1979) 625–642;
M. Berry, "Some quantum-to-classical asymptotics", in *Les Houches Lecture Series LII (1989)*, M-J. Giannoni, A. Voros and J. Zinn-Justin (eds.), (North-Holland, Amsterdam, 1991), pp. 251–304

BB79. M. Berry and N. Balazs, *Am J Phys* **47** (1979) 264–267

BCG97. M. Bertelson, M. Cahen, and S. Gutt, *Class Quant Grav* **14** (1997) A93–A107.

Ber02. P. Bertet *et al.*, *Phys Rev Lett* **89** (2002) 200402

BGR91. I. Bialynicki-Birula, P. Górnicki, and J. Rafelski, *Phys Rev* **D44** (1991) 1825–1835

BP96. B. Biegel and J. Plummer, *Phys Rev* **B54** (1996) 8070–8082

BHS02. M. Bienert, F. Haug, W. Schleich, and M. Raizen, *Phys Rev Lett* **89** (2002) 050403

BHP02. P. Bianucci *et al.*, *Phys Lett* **A297** (2002) 353–358

BV94. R. Bishop and A. Vourdas, *Phys Rev* **A50** (1994) 4488–4501

Blo40. D. Blokhintsev, *J Phys [of the USSR]* **2** (1940) 71–74

BTU93. O. Bohigas, S. Tomsovic, and D. Ullmo, *Phys Rep* **223** (1993) 43–133

Bon84. J. Gracia-Bondía, *Phys Rev* **A30** (1984) 691–697

Bor96. M. Bordemann, M. Brischle, C. Emmrich, and S. Waldmann, *Lett Math Phys* **36** (1996) 357–371

BCT82. E. Braaten, T. Curtright, and C. Thorn, *Phys Lett* **B118** (1982) 115

BM94. A. Bracken and G. Melloy, *J Phys* **A27** (1994) 2197–2211

BDW99. A. Bracken, H. Doebner, and J. Wood, *Phys Rev Lett* **83** (1999) 3758–3761;
A. Bracken, D. Ellinas, and J. Wood *J Phys* **A36** (2003) L297–L305;
J. Wood and A. Bracken *J Math Phys* **46** (2005) 042103

BCW02. A. Bracken, G. Cassinelli, and J. Wood, *J Phys* **A36** (2003) 1033–1057 [math-ph/0211001]

Bra03. A. Bracken, *J Phys* **A36** (2003) L329-L335;
A. Bracken and J. Wood, *Phys Rev* **A73** (2006) 012104

BR93. G. Braunss and D. Rompf, *J Phys* **A26** (1993) 4107–4116

Bra94. G. Braunss, *J Math Phys* **35** (1994) 2045–2056

BC62. W. Brittin and W. Chappell, *Rev Mod Phys* **34** (1962) 620–627

BD98. D. Brown and P. Danielewicz, *Phys Rev* **D58** (1998) 094003

Bou99. A. Boulkhemair, *J Funct Anal* **165** (1999) 173–204

BAD96. V. Bužek, G. Adam, and G. Drobný, *Ann Phys (NY)* **245** (1996) 37–97

CC03. A. Cafarella, C. Corianò, and M. Guzzi, *JHEP* **11** (2003) 059

CGR90. M. Cahen, S. Gutt, and J. Rawnsley, *J Geom Phys* **7** (1990) 45–62

CL83. A. Caldeira and A. Leggett, *Physica* **121A** (1983) 587–616

CBJ07. E. Cancellieri, P. Bordone, and C. Jacoboni, *Phys Rev* **B76** (2007) 214301

CZ83. P. Carruthers and F. Zachariasen, *Rev Mod Phys* **55** (1983) 245–285

Car76. N. Cartwright, *Physica* **83A** (1976) 210–213

CdD04. A. Carvalho, R. de Matos Filho, and L. Davidovich, *Phys Rev* **E70** (2004) 026211

Cas91. M. Casas, H. Krivine, and J. Martorell, *Eur J Phys* (1991) 105–111

Cas00. L. Castellani, *Class Quant Grav* **17** (2000) 3377–3402 [hep-th/0005210]

CSA09. S. Chaudhury, A. Smith, B. Anderson, S. Ghose, and P. Jessen, *Nature* **461** (2009) 768–771

CH87. L. Chetouani and T. Hammann, *J Math Phys* **28** (1987) 598–604

CH86. L. Chetouani and T. Hammann, *Nuov Cim* **B92** (1986) 106–120

CV98. S. Chountasis and A. Vourdas, *Phys Rev* **A58** (1998) 1794–1798

CL03. Y-J. Chun and H-W. Lee, *Ann Phys (NY)* **307** (2003) 438–451

CKTM07. W. Coffey, Yu. Kalmykov, S. Titov, and B. Mulligan, *Phys Rev* **E75** (2007) 041117; *ibid.* **E78** (2008) 031114

Coh95. L. Cohen, *Time-Frequency Analysis* (Prentice Hall PTR, Englewood Cliffs, 1995)

Coh66. L. Cohen, *J Math Phys* **7** (1966) 781

Coh76. L. Cohen, *J Math Phys* **17** (1976) 1863

Con37. E. Condon, *Proc Nat Acad Sci USA* **23** (1937) 158–164

CS75. F. Cooper and D. Sharp, *Phys Rev* **D12** (1975) 1123–1131;
R. Hakim and J. Heyvaerts, *Phys Rev* **A18** (1978) 1250–1260

CPP01. H. García-Compeán, J. Plebanski, M. Przanowski, and F. Turrubiates, *Int J Mod Phys* **A16** (2001) 2533–2558

CPP02. H. García-Compeán, J. Plebanski, M. Przanowski, and F. Turrubiates, *J Phys* **A35** (2002) 4301–4320

CGB91. G. Cristóbal, C. Gonzalo, and J. Bescós, *Adv Electron Electron Phys* **80** (1991) 309–397

CG92. T. Curtright and G. Ghandour, in *Quantum Field Theory, Statistical Mechanics, Quantum Groups and Topology, Coral Gables 1991 Proceedings*, T. Curtright *et al.* (eds.) (World Scientific, 1992), pp. 333–344 [hep-th/9503080]

CUZ01. T. Curtright, T. Uematsu, and C. Zachos, *J Math Phys* **42** (2001) 2396–2415 [hep-th/0011137]

CV07. T. Curtright and A. Veitia, *J Math Phys* **48** (2007) 102112 [quant-ph/0701006]

CZ99. T. Curtright and C. Zachos, *J Phys* **A32** (1999) 771–779

CFZ98. T. Curtright, D. Fairlie, and C. Zachos, *Phys Rev* **D58** (1998) 025002

CFZm98. T. Curtright, D. Fairlie, and C. Zachos, "Matrix Membranes and Integrability" in *Supersymmetry and Integrable Models*, H. Aratyn *et al.* (eds.), Lecture Notes in Physics 502 (Springer-Verlag, Heidelberg, 1998), pp. 183–196 [hep-th/9709042]

CZ01. T. Curtright and C. Zachos, *Mod Phys Lett* **A16** (2001) 2381–2385

CZ02. T. Curtright and C. Zachos, *New J Phys* **4** (2002) 1.1–1.16 [hep-th/0205063]

CZ12. T. Curtright and C. Zachos, *Asia Paci Phys Newslett* **1** (2012) 37–46

DS82. J. Dahl and M. Springborg, *Mol Phys* **47** (1982) 1001–1019; *Phys Rev* **A36** (1988) 1050–1062; *Phys Rev* **A59** (1999) 4099–4100; *J Chem Phys* **88** (1988) 4535–4547

Dah01. J. P. Dahl, *Adv Quant Chem* **39** (2001) 1–18

DS02. J. P. Dahl, and W. Schleich, *Phys Rev* **A65** (2002) 022109

Dah83. J. Dahl, in *Energy Storage and Redistribution*, J. Hinze (ed.) (Plenum Press, New York, 1983), pp. 557–571

DG80. I. Daubechies and A. Grossmann, *J Math Phys* **21** (1980) 2080–2090;
I. Daubechies, A. Grossmann, and J. Reignier, *J Math Phys* **24** (1983) 239–254

DK85. I. Daubechies and J. Klauder, *J Math Phys* **26** (1985) 2239–2256

DG02. E. Davis and G. Ghandour, *J Phys* **35** (2002) 5875–5891 [quant-ph/9905002]

deA98. A. M. Ozorio de Almeida, *Phys Rep* **295** (1998) 265–342

deB67. N. G. de Bruijn, "Uncertainty Principles in Fourier Analysis" in *Inequalities*, O. Shisha (ed.) (Academic Press, New York, 1967), pp. 57–71

deB73. N. G. de Bruijn, *Nieuw Arch Wiskd, III. Ser* **21** (1973) 205–280

deG74. S. de Groot, *La Transformation de Weyl et la fonction de Wigner* (Presses de l' Université de Montreal, 1974);
S. de Groot and L. Suttorp, *Foundations of Electrodynamics* (North Holland, Amsterdam, 1972)

Dek77. H. Dekker, *Phys Rev* **A16** (1977) 2126–2134

DBB02. L. Demeio, L. Barletti, A. Bertoni, P. Bordone, and C. Jacoboni, *Physica* **B314** (2002) 104–107;
L. Demeio, P. Bordone, and C. Jacoboni, *Trans Th Stat Phys* **34** (2006) 1;
O. Morandi and L. Demeio, *Trans Th Stat Phys* **37** (2008) 437

dMF06. C. de Morisson Faria and A. Fring, *Czech J Phys* **56** (2006) 899 [quant-ph/0607154]

DV97. T. Dereli and A. Vercin, *J Math Phys* **38** (1997) 5515–5530 [quant-ph/9707040] deW83. M. de Wilde and P. Lecomte, *Lett Math Phys* **7** (1983) 487

DP01. N. Dias and J. Prata, *J Math Phys* **42** (2001) 5565–5579

DO85. R. Dickman and R. O'Connell, *Phys Rev* **B32** (1985) 471–473

Dir25. P. Dirac, *Proc Roy Soc Lond* **A109** (1925) 642–653

Dir30. P. Dirac, *Proc Camb Phil Soc* **26** (1930) 376–385

Dir33. P. Dirac, *Phys Z Sowjetunion* **3** (1933) 64–72

Dir45. P. Dirac, *Rev Mod Phys* **17** (1945) 195–199

DKM88. R. Dirl, P. Kasperkovitz, and M. Moshinsky, *J Phys* **A21** (1988) 1835–1846

Dit90. J. Dito, *Lett Math Phys* **20** (1990) 125–134; *J Math Phys* **33** (1992) 791–801

DVS06. T. Dittrich, C. Viviescas, and L. Sandoval, *Phys Rev Lett* **96** (2006) 070403

DM86. V. Dodonov and V. Man'ko, *Physica* **137A** (1986) 306–316

DN01. M. Douglas and N. Nekrasov, *Rev Mod Phys* **73** (2001) 977–1029;
R. Szabo, *Phys Rep* **378** (2003) 207–299

DHS00. D. Dubin, M. Hennings, and T. Smith, *Mathematical Aspects of Weyl Quantization and Phase* (World Scientific, Singapore, 2000)

Dun95. T. Dunne *et al.*, *Phys Rev Lett* **74** (1995) 884–887

Dun88. G. Dunne, *J Phys* **A21** (1988) 2321–2335

EGV89. R. Estrada, J. Gracia-Bondía, and J. Várilly, *J Math Phys* **30** (1989) 2789–2796

Fai64. D. Fairlie, *Proc Camb Phil Soc* **60** (1964) 581–586

FFZ89. D. Fairlie and C. Zachos, *Phys Lett* **B224** (1989) 101–107;

D. Fairlie, P. Fletcher, and C. Zachos, *J Math Phys* **31** (1990) 1088–1094

FM91. D. Fairlie and C. Manogue, *J Phys* **A24** (1991) 3807–3815

Fan03. A. Fannjiang, *Commun Math Phys* **254** (2005) 289–322 [math-ph/0304024]

Fan57. U. Fano, *Rev Mod Phys* **29** (1957) 74–93

FBA96. A. Farini, S. Boccaletti, and F. Arecchi, *Phys Rev* **E53** (1996) 4447–4450

FZ01. A. Fedorova and M. Zeitlin, in *PAC2001 Proceedings*, P. Lucas and S. Webber (eds.), (IEEE, Piscataway, NJ, 2001) 1814–1816 [physics/0106005];
A. Fedorova and M. Zeitlin, *18th Advanced ICFA Beam Dynamics Workshop on Quantum Aspects of Beam Physics: Capri, 2000*, P. Chen (ed.) (World Scientific, River Edge, NJ, 2002) 539–550 [physics/0101006]

Fed94. B. Fedosov, *J Diff Geom* **40** (1994) 213–238

Fey87. R. Feynman, "Negative Probability" in *Essays in Honor of David Bohm*, B. Hiley and F. Peat (eds.) (Routledge and Kegan Paul, London, 1987) 235–248

FM03. S. Filippas and G. Makrakis *Multiscale Mod Simul* **1** (2003) 674–710

Fil96. T. Filk, *Phys Lett* **B376** (1996) 53–58

FLM98. W. Fischer, H. Leschke, and P. Müller, and P. Müller, *Annalen Phys* **7** (1998) 59–100 [quant-ph/9807065]; *Phys Rev Lett* **73** (1994) 1578–1581

Fla99. P. Flandrin, *Time-Frequency/Time-scale Analysis (Wavelet Analysis and Its Applications)* (Academic Press, San Diego, 1999)

FLS76. M. Flato, A. Lichnerowicz, and D. Sternheimer, *J Math Phys* **17** (1976) 1754

Fle90. P. Fletcher, *Phys Lett* **B248** (1990) 323–328

Fol89. G. Folland *Harmonic Analysis in Phase Space* (Princeton University Press, Princeton, 1989)

FO01. G. Ford and R. O'Connell, *Phys Rev* **D64** (2001) 105020

FO05. G. Ford and R. O'Connell, *Ann Phys (NY)* **319** (2005) 348

FO07. G. Ford and R. O'Connell, *Phys Rev* **A76** (2007) 042122

FO10. G. Ford, Y. Gao, and R. O'Connell, *Opt Commun* **283** (2010) 831

FO11. G. Ford and R. O'Connell, *J Comput Theor Nanosci* **8** (2011) 1–7

Fra00. A. Frank, A. Rivera, and K. Wolf, *Phys Rev* **A61** (2000) 054102

Fre87. W. Frensley, *Phys Rev* **B36** (1987) 1570–1578; W. Frensley, *Rev Mod Phys* **62** (1990) 745–791

FMS00. O. Friesch, I. Marzoli, and W. Schleich, *New J Phys* **2** (2000) 4.1–4.11

Gad95. M. Gadella, *Fortschr Phys* **43** (1995) 3, 229–264

GM80. G. García-Calderón and M. Moshinsky, *J Phys* **A13** (1980) L185

GK94. B. Garraway and P. Knight, *Phys Rev* **A50** (1994) 2548–2563

Gat07. O. Gat, *J Phys* **A40** (2007) F911–F920

GF91. I. Gelfand and D. Fairlie, *Commun Math Phys* **136** (1991) 487–500

GH93. M. Gell-Mann and Hartle, *Phys Rev* **D47** (1993) 3345–3382

GB03. J. Gong and P. Brumer, *Phys Rev* **A68** (2003) 062103

GLL10. M. Gorbunov, K. Lauritsen, and S. Leroy, *Radio Sci* **45** (2010) RS6011

Got99. M. Gotay, *J Math Phys* **40** (1999) 2107–2116

GR94. E. Gozzi and M. Reuter, *Int J Mod Phys* **A9** (1994) 5801–5820

GHSS05. B. Greenbaum, S. Habib, K. Shizume, and B. Sundaram, *Chaos* **15** (2005) 033302

Gro01. K. Gröchenig, *Foundations of Time-Frequency Analysis* (Birkhäuser, Boston, 2001)

Gro46. H. Groenewold, *Physica* **12** (1946) 405–460

GLS68. A. Grossmann, G. Loupias, and E. Stein, *Ann Inst Fourier* **18** (1968) 343–368

Gro76. A. Grossmann, *Commun Math Phys* **48** (1976) 191–194

Haa10. F. Haake, *Quantum Signatures of Chaos*, Springer Series in Synergetics, vol. 54 (Springer, 2010)

Hab90. S. Habib, *Phys Rev* **D42** (1990) 2566–2576;
S. Habib and R. Laflamme, *Phys Rev* **D42** (1990) 4056–4065

Hak99. T. Hakioğlu, *J Phys* **A32** (1999) 4111–4130; *J Opt Soc Am* **A17** (2000) 2411–2421; T. Hakioğlu and A. Dragt, *J Phys* **A34** (2002) 6603–6615; T. Hakioğlu, A. Teğmen, and B. Demircioğlu, *Phys Lett* **A360** (2007) 501–506; T. Dereli, T. Hakioğlu, and A. Teğmen, *Int J Mod Phys* **A24** (2009) 4573–4587.

HY96. J. Halliwell and T. Yu, *Phys Rev* **D53** (1996) 2012–2019

HKN88. D. Han, Y. Kim, and M. Noz, *Phys Rev* **A37** (1988) 807–814;
Y. Kim and E. Wigner, *ibid* **A38** (1988) 1159–1167; *ibid* **A36** (1987) 1293–1297

Han84. F. Hansen, *Rep Math Phys* **19** (1984) 361–381

Har01. J. Harvey, "Komaba Lectures on Noncommutative Solitons and D-branes" [hep-th/0102076]

HS02. A. Hatzinikitas and A. Smyrnakis, *J Math Phys* **A43** (2002) 113–125

Hel76. E. Heller, *J Chem Phys* **65** (1976) 1289–1298; *ibid* **67** (1977) 3339–3351

HSD95. M. Hennings, T. Smith, and D. Dubin, *J Phys* **A28** (1995) 6779–6807; *ibid* 6809–6856

Hie82. J. Hietarinta, *Phys Rev* **D25** (1982) 2103–2117

Hie84. J. Hietarinta, *J Math Phys* **25** (1984) 1833–1840.

HOS84. M. Hillery, R. O'Connell, M. Scully, and E. Wigner, *Phys Rep* **106** (1984) 121–167

HH02. A. Hirschfeld and P. Henselder, *Am J Phys* **70** (2002) 537–547

HP03. M. Horvat and T. Prosen, *J Phys* **A36** (2003) 4015–4034

Hor79. L. Hörmander, *Commun Pure Appl Math* **32** (1979) 359–443;
*The Analysis of Linear Partial Differential Operators II & III* (Springer Verlag, Berlin-Heidelberg, 1985)

HL99. X-G. Hu and Q-S. Li, *J Phys* **A32** (1999) 139–146

Hud74. R. Hudson, *Rep Math Phys* **6** (1974) 249–252

HMS98. M. Hug, C. Menke, and W. Schleich, *Phys Rev* **A57** (1998) 3188–3205; *ibid* 3206–3224

HW80. J. Hutchinson and R. Wyatt, *Chem Phys Lett* **72** (1980) 378–384

Hus40. K. Husimi, *Proc Phys Math Soc Jpn* **22** (1940) 264

Iag69. D. Iagolnitzer, *J Math Phys* **10** (1969) 1241–1264

Imr67. K. İmre, K. Özizmir, M. Rosenbaum, and P. Zweifel, *J Math Phys* **8** (1967) 1097–1108

IZ51. J. Irving and R. Zwanzig, *J Chem Phys* **19** (1951) 1173–1180

JBM03. C. Jacoboni, R. Brunetti, and S. Monastra, *Phys Rev* **B68** (2003) 125205;
C. Jacoboni *et al.*, *Rep Prog Phys* **67** (2004) 1033–1071;
P. Bordone, C. Jacoboni, *et al.*, *Phys Rev* **B59** (1999) 3060–3069

Jan78. B. Jancovici, *Physica* **A91** (1978) 152–160;
A. Alastuey and B. Jancovici, *ibid* **A102** (1980) 327–343

Jan84. A. Janssen, *J Math Phys* **25** (1984) 2240–2252

JS02. Y. Japha and B. Segev, *Phys Rev* **A65** (2002) 063411

JVS87. J. Javanainen, S. Varró, and O. Serimaa, *Phys Rev* **A35** (1987) 2791–2805;
*ibid* **A33** (1986) 2913–2927

JG93. K. Jensen and A. Ganguly, *J Appl Phys* **73** (1993) 4409–4427

JN90. J. Jensen and Q. Niu, *Phys Rev* **A42** (1990) 2513–2519

JY98. A. Jevicki and T. Yoneya, *Nucl Phys* **B535** (1998) 335

JD99. A. Joshi and H-T. Dung, *Mod Phys Lett* **B13** (1999) 143–152

Kar98. M. Karasev, *Diff Geom Appl* **9** (1998) 89–134

KO00. M. Karasev and T. Osborn, *J Math Phys* **43** (2002) 756–788 [quant-ph/0002041];
*J Phys* **A37** (2004) 2345–2363 [quant-ph/0311053];
*ibid* **A38** (2005) 8549–8578

KP94. P. Kasperkovitz and M. Peev, *Ann Phys (NY)* **230** (1994) 21–51

KZZ02. Z. Karkuszewski, J. Zakrzewski, and W. Zurek, *Phys Rev* **A65** (2002) 042113;
Z. Karkuszewski, C. Jarzynski, and W. Zurek *Phys Rev Lett* **89** (2002) 170405

KT07. B. Kaynak and T. Turgut, *J Math Phys* **48** (2007) 113501

KJ99. C. Kiefer and E. Joos, in *Quantum Future*, P. Blanchard and A. Jadczyk (eds.) (Springer-Verlag, Berlin, 1999), pp. 105–128 [quant-ph/9803052];
L. Diósi and C. Kiefer, *J Phys* **A35** (2002) 2675–2683;
E. Joos, H. Zeh, C. Kiefer, D. Giulini, J. Kupsch, and I-O. Stamatescu, *Decoherence and the Appearance of a Classical World in Quantum Theory* (Springer Verlag, Heidelberg, 2003)

KL99. J-H. Kim and H-W. Lee, *Can J Phys* **77** (1999) 411–425

KL01. K-Y. Kim and B. Lee, *Phys Rev* **B64** (2001) 115304

KN91. Y. Kim and M. Noz, *Phase Space Picture of Quantum Mechanics*, Lecture Notes in Physics, vol. 40 (World Scientific, Singapore, 1991)

KW90. Y. Kim and E. Wigner, *Am J Phys* **58** (1990) 439–448

KW87. Y. Kim and E. Wigner, *Phys Rev* **A36** (1987) 1293; *ibid* **A38** (1988) 1159

Kir33. J. Kirkwood, *Phys Rev* **44** (1933) 31–37;
(E) *ibid* **45** (1934) 116–117;
G. Uhlenbeck and L. Gropper, *Phys Rev* **41** (1932) 79–90

Kis01. I. Kishimoto, *JHEP* **0103** (2001) 25

KL92. S. Klimek and A. Lesniewski, *Commun Math Phys* **146** (1992) 103–122

KKFR89. N. Kluksdahl, A. Kriman, D. Ferry, and C. Ringhofer, *Phys Rev* **B39** (1989) 7720–7735

Kol96. A. Kolovsky, *Phys Rev Lett* **76** (1996) 340–343

KS02. A. Konechny and A. Schwarz, *Phys Rep* **360** (2002) 353–465

KL94. H. Konno and P. Lomdahl, *J Phys Soc Jpn* **63** (1994) 3967–3973

Kon97. M. Kontsevich, *Lett Math Phys* **66** (2003) [q-alg/9709040]; *ibid* **48** (1999) 35–72 [math.QA/9904055]

KB81. H. Korsch and M. Berry, *Physica* **3D** (1981) 627–636

KW05. S. Kryukov and M. Walton, *Ann Phys (NY)* **317** (2005) 474–491; *Can J Phys* **84** (2006) 557–563

Kub64. R. Kubo, *J Phys Soc Jpn* **19** (1964) 2127–2139

Kun67. W. Kundt, *Z Nat Forsch* **A22** (1967) 1333–1336

KPM97. C. Kurtsiefer, T. Pfau, and J. Mlynek, *Nature* **386** (1997) 150

Kut72. J. Kutzner, *Phys Lett* **A41** (1972) 475–476; *Zeit f Phys* **A259** (1973) 177–188

Les84. B. Lesche, *Phys Rev* **D29** (1984) 2270–2274

Lea68. B. Leaf, *J Math Phys* **9** (1968) 65–72; *ibid* **9** (1968) 769–781

LS82. H-W. Lee and M. Scully, *J Chem Phys* **77** (1982) 4604–4610

Lee95. H-W. Lee, *Phys Rep* **259** (1995) 147–211

Lei96. D. Leibfried *et al.*, *Phys Rev Lett* **77** (1996) 4281

LPM98. D. Leibfried, T. Pfau, and C. Monroe, *Phys Today* **51** (1998) 22–28

Leo97. U. Leonhardt, *Measuring the Quantum State of Light* (Cambridge University Press, Cambridge, 1997)

LF01. M. Levanda and V. Fleurov, *Ann Phys (NY)* **292** (2001) 199–231

LF94. M. Levanda and V. Fleurov, *J Phys: Condens Matt* **6** (1994) 7889–7908

LSU02. B. Lev, A. Semenov, and C. Usenko, *J Russ Laser Res* **23** (2002) 347–368 [quant-ph/0112146]

Lie90. E. Lieb, *J Math Phys* **31** (1990) 594–599

Lit86. R. Littlejohn, *Phys Rep* **138** (1986) 193

Lou96. P. Loughlin (ed.), *Special Issue on Time Frequency Analysis, Proceeding of the IEEE* **84**(9) (2001)

Lut96. L. Lutterbach and L. Davidovich, *Phys Rev Lett* **78** (1997) 2547–2550

Lvo01. A. Lvovsky *et al.*, *Phys Rev Lett* **87** (2001) 050402

Mah87. G. Mahan, *Phys Rep* **145** (1987) 251

MNV08. R. Maia, F. Nicasio, R. Vallejos, and F. Toscano, *Phys Rev Lett* **100** (2008) 184102

MS95. M. Mallalieu and C. Stroud, *Phys Rev* **A51** (1995) 1827–1835

MM84. J. Martorell and E. Moya, *Ann Phys (NY)* **158** 1–30

MMT96. S. Mancini, V. Man'ko, and P. Tombesi, *Phys Lett* **A213** (1996) 1–6

MMM01. O. Man'ko, V. Man'ko, and G. Marmo, in *Quantum Theory and Symmetries, Krakow 2001 Proceedings*, E. Kapuscik and A. Horzela (eds.) (World Scientific, 2002) [quant-ph/0112112];
G. Amosov, Ya. Korennoy, and V. Manko, *Phys Rev* **A85** (2012) 052119

MP04. M. Mantoiu and R. Purice, *J Math Phys* **45** (2004) 1394–1416

Mar91. M. Marinov, *Phys Lett* **A153** (1991) 5–11

MS96. M. Marinov and B. Segev, *Phys Rev* **A54** (1996) 4752–4762;

B. Segev, in *Michael Marinov Memorial Volume: Multiple Facets of Quantization and Supersymmetry*, M Olshanetsky and A Vainstein (eds.) (Worlds Scientific, 2002)

MMP94. P. Markowich, N. Mauser, and F. Poupaud, *J Math Phys* **35** (1995) 1066–1094

McD88. S. McDonald, *Phys Rep* **158** (1988) 337–416

McC32. N. McCoy, *Proc Nat Acad Sci USA* **19** (1932) 674

MOT98. B. McQuarrie, T. Osborn, and G Tabisz, *Phys Rev* **A58** (1998) 2944–2960;
T. Osborn, M. Kondrat'eva, G. Tabisz, and B. McQuarrie, *J Phys* **A32** (1999) 4149–4169

MMM11. C. C. Meaney, R. McKenzie, and G. Milburn, *Phys Rev* **E83** (2011) 056202

MH97. W. Mecklenbräuker and F. Hlawatsch (eds.), *The Wigner Distribution* (Elsevier, Amsterdam, 1997);
G. Matz and F. Hlawatsch, *J Math Phys* **39** (1998) 4041–4069.

MPS02. C. Miquel, J. P. Paz, and M. Saraceno, *Phys Rev* **A65** (2002) 062309

Mon94. T. Monteiro, *J Phys* **A27** (1994) 787–800

Mor09. O. Morandi, *Phys Rev* **B80** (2009) 024301;
O. Morandi and F. Schuerrer *J Phys* **A44** (2011) 265301

Mor86. C. Moreno, *Lett Math Phys* **11** (1986) 361–372

Mos05. A. Mostafazadeh, *Proc 3rd Int Workshop Pseudo-Hermitian Hamiltonians in Quantum Physics*, June 20–22, 2005, Koc University, Istanbul, Turkey [quant-ph/0508214]

Moy49. J. Moyal, *Proc Camb Phil Soc* **45** (1949) 99–124

Moy06. A. Moyal, *Maverick Mathematician*, (ANU E Press, 2006), online: http://epress.anu.edu.au/maverick/mobile_devices/

MM94. S. Mrówczyński and B. Müller, *Phys Rev* **D50** (1994) 7542–7552

MLD86. W. Mückenheim, G. Ludwig, C. Dewdney, *et al.*, *Phys Rep* **133** (1986) 337–401

Mue99. M. Müller, *J Phys* **A32** (1999) 1035–1052

Na97. H. Nachbagauer [hep-th/9703105]

NO86. F. Narcowich and R. O'Connell, *Phys Rev* **A34** (1986) 1–6;
F. Narkowich *J Math Phys* **28** (1987) 2873–2882

Neu31. J. von Neumann, *Math Ann* **104** (1931) 570–578

NH08. S. Nimmrichter and K. Hornberger, *Phys Rev* **A78** (2008) 023612

OW81. R. O'Connell and E. Wigner, *Phys Lett* **85A** (1981) 121–126

OW84. R. O'Connell and E. Wigner, *Phys Rev* **A30** (1984) 2613

OC03. R. O'Connell, *J Opt* **B5** (2003) S349–S359;
M. Murakami, G. Ford, and R. O'Connell, *Laser Phys* **13** (2003) 180–183

OBB95. T. Opatrný, V. Bužek, J. Bajer, and G. Drobný, *Phys Rev* **A52** (1995) 2419–2428; *Phys Rev* **A53** (1996) 3822–3835

OR57. I. Oppenheim and J. Ross, *Phys Rev* **107** (1957) 28–32

OM95. T. Osborn and F. Molzahn, *Ann Phys (NY)* **241** (1995) 79–127

Pei33. R. Peierls, *Z Phys* **80** (1933) 763

PT99. M. Przanowski and J. Tosiek, *Act Phys Pol* **B30** (1999) 179–201

Pul06. M. Pulvirenti, *J Math Phys* **47** (2006) 052103

QC96. S. Qian and D. Chen, *Joint Time-Frequency Analysis* (Prentice Hall PTR, Upper Saddle River, NJ, 1996)

Raj83. A. Rajagopal, *Phys Rev* **A27** (1983) 558–561

Raj02. S. G. Rajeev, in *Proc 70th Meeting of Mathematicians and Physicists* June 2002, Strassbourg, V. Turaev and T. Wurzbacher (eds.), [hep-th/0210179]

Ram04. J. Rammer, *Quantum Transport Theory Frontiers in Physics*, (Westview Press, Boulder, 2004)

Ran66. B. Rankin, *Phys Rev* **141** (1966) 1223–1230

Raz96. M. Razavy, *Phys Lett* **A212** (1996) 119–122

RT00. N. Reshetikhin and L. Takhtajan, *Am Math Soc Trans* **201** (2000) 257–276 [math.QA/9907171]

Rie89. M. Rieffel, *Commun Math Phys* **122** (1989) 531–562

RA99. A. Rivas and A. O. de Almeida, *Ann Phys (NY)* **276** (1999) 223–256

Rob93. S. Robinson, *J Math Phys* **34** (1993) 2185–2205

RO92. C. Roger and V. Ovsienko, *Russ Math Surv* **47** (1992) 135–191

Roy77. A. Royer, *Phys Rev* **A15** (1977) 449–450

RG00. M. Ruzzi and D. Galetti, *J Phys* **A33** (2000) 1065–1082;
D. Galetti and A. de Toledo Piza, *Physica* **149A** (1988) 267–282

Sam00. J. Samson, *J Phys* **A33** (2000) 5219–5229 [quant-ph/0006021]

Sch88. W. Schleich, H. Walther, and J. A. Wheeler, *Found Phys* **18** (1988) 953–968

Sch02. W. Schleich, *Quantum Optics in Phase Space* (Wiley-VCH, 2002)

Sch69. J. Schipper, *Phys Rev* **184** (1969) 1283–1302

SG05. F. Scholtz and H. Geyer, *Phys Lett* **B634** (2006) 84, [quant-ph/0512055]

SG06. F. Scholtz and H. Geyer, *J Phys* **39** (2006) 10189–10205 [quant-ph/0602187]

SGH92. F. Scholtz, H. Geyer, and F. Hahne, *Ann Phys* **213** (1992) 74–101

SchN59. K. Schram and B. Nijboer, *Physica* **25** (1959) 733–741

SW99. N. Seiberg and E. Witten, *JHEP* **9909** (1999) 32

SST00. N. Seiberg, L. Susskind, and N. Toumbas, *JHEP* **0006** (2000) 44 [hep-th/0005015]

SS02. A. Sergeev and B. Segev, *J Phys* **A35** (2002) 1769–1789;
B. Segev, *J Opt* **B5** (2003) S381–S387

Sha79. P. Sharan, *Phys Rev* **D20** (1979) 414–418

SLC11. S. Shao, T. Lu, and W. Cai, *Commun Comput Phys* **9** (2011) 711–739

She59. J. Shewell, *Am J Phys* **27** (1959) 16–21

SRF03. L. Shifren, C. Ringhofer, and D. Ferry, *IEEE Transactions on Electron Devices* **50** (2003) 769–773

Shi79. Yu. Shirokov, *Sov J Part Nucl* **10** (1979) 1–18

SP81. S. Shlomo and M. Prakash, *Nucl Phys* **A357** (1981) 157

SM00. R. Simon and N. Mukunda, *J Opt Soc Am* **A17** (2000) 2440–2463

Smi93. D. Smithey *et al.*, *Phys Rev Lett* **70** (1993) 1244–1247

Sny80. J. Snygg, *Am J Phys* **48** (1980) 964–970

Son09. W. Son *et al.*, *Phys Rev Lett* **102** (2009) 110404

Ste80. S. Stenholm, *Eur J Phys* **1** (1980) 244–248

SKR13. O. Steuernagel, D. Kakofengitis, and G. Ritter, *Phys Rev Lett* **110** (2013) 030401

Str57. R. Stratonovich, *Sov Phys JETP* **4** (1957) 891–898

Syl82. J. Sylvester, *Johns Hopkins University Circulars* **I** (1882) 241–242; *ibid* **II** (1883) 46; *ibid* **III** (1884) 7–9. Summarized in *The Collected Mathematics Papers of James Joseph Sylvester* (Cambridge University Press, 1909), vol. III

Tak54. T. Takabayasi, *Prog Theor Phys* **11** (1954) 341–373

Tak89. K. Takahashi, *Prog Theor Phys Suppl* **98** (1989) 109–156

Tat83. V. Tatarskii, *Sov Phys Usp* **26** (1983) 311

Tay01. W. Taylor, *Rev Mod Phys* **73** (2001) 419

Ter37. Y. P. Terletsky, *Zh Eksp Teor Fiz*, **7** (1937) 1290–1298;
D. Rivier, *Phys Rev* **83** (1951) 862–863

TGS05. M. Terraneo, B. Georgeot, and D. L. Shepelyansky, *Phys Rev* **E71** (2005) 066215

TW03. C. Trahan and R. Wyatt, *J Chem Phys* **119** (2003) 7017–7029

TKS83. M. Toda, R. Kubo, and N. Saitô, *Statistical Physics I: Equilibrium Statistical Mechanics* (Springer, Berlin, 1983)

Tod12. I. Todorov, *Bulg J Phys* **39** (2012) 107–149 [arXiv:1206.3116]

TZM96. Go. Torres-Vega, A. Zúñiga-Segundo, and J. Morales-Guzmán, *Phys Rev* **A53** (1996) 3792–3797

TA99. F. Toscano and A. O. de Almeida, *J Phys* **A32** (1999) 6321–6346

TD97. C. Tzanakis and A. Dimakis, *J Phys* **A30** (1997) 4857–4866

Unt79. A. Unterberger, *Ann Inst Fourier* **29** (1979) 201–221
VG89. J. Várilly and J. Gracia-Bondía, *Ann Phys (NY)* **190** (1989) 107–148
Vill48. J. Ville, *Câbles et Transmissions* **2** (1948) 61–74
VMdG61. J. Vlieger, P. Mazur, and S. de Groot, *Physica* **27** (1961) 353–372; 957–960; 974–978
Vog89. K. Vogel and H. Risken, *Phys Rev* **A40** (1989) 2847–2849
Vor89. A. Voros, *Phys Rev* **A40** (1989) 6814–6825
Vo78. A. Voros, *J Funct Anal* **29** (1978) 104–132; *Ann Inst H Poincaré* **24** (1976) 31–90; *ibid* **26** (1977) 343–403; B. Grammaticos and A. Voros, *Ann Phys (NY)* **123** (1979) 359–380
Vey75. J. Vey, *Comment Math Helv* **50** (1975) 421–454
vH51. L. van Hove, *Mem Acad Roy Belgique* **26** (1951) 61–102
vZy12. B. van Zyl, *J Phys* **A45** (2012) 315302
WO87. L. Wang and R. O'Connell, *Physica* **144A** (1987) 201–210
WO88. L. Wang and R. O'Connell, *Found Phys* **18** (1988) 1023–1033
Wer95. R. Werner [quant-ph/9504016]
Wey27. H. Weyl, *Z Phys* **46** (1927) 1–33;
H. Weyl, *The Theory of Groups and Quantum Mechanics* (Dover, New York, 1931)
WH99. U. Wiedemann and U. Heinz, *Phys Rep* **319** (1999) 145–230
Wig32. E. Wigner, *Phys Rev* **40** (1932) 749–759
Wis97. H. Wiseman *et al.*, *Phys Rev* **A56** (1997) 55–75
Wok97. W. Wokurek, in *Proc ICASSP'97* (Munich, 1997), pp. 1435–1438

Wo98. M-W. Wong, *Weyl Transforms* (Springer-Verlag, Berlin, 1998)

Wo82. C-Y. Wong, *Phys Rev* **C25** (1982) 1460–1475

Wo02. C-Y. Wong, *J Opt* **B5** (2003) S420–S428 [quant-ph/0210112]

Woo87. W. Wootters, *Ann Phys (NY)* **176** (1987) 1–21;
K. Gibbons, M. Hoffman, and W. Wootters, *Phys Rev* **A70** 062101;
W. Wootters, *IBM J Res Dev* **48** (2004) 99–110 [quant-ph/0306135]

WL10. X. Wu and T. Liu, *J Geophys Eng* **7** (2010) 126

XA96. A. Xavier Jr. and M. de Aguiar, *Ann Phys (NY)* **252** (1996) 458–478

Xu98. P. Xu, *Commun Math Phys* **197** (1998) 167–197

Yv46. J. Yvon, *Comput Rend Acad Sci* **223** (1946) 347–349

Yo89. T. Yoneya, *Mod Phys Lett* **A4** (1989) 1587

Zac00. C. Zachos, *J Math Phys* **41** (2000) 5129–5134 [hep-th/9912238];
C. Zachos, "A Survey of Star Product Geometry" in *Integrable Hierarchies and Modern Physical Theories*, H. Aratyn and A Sorin (eds.), NATO Science Series II, vol. 18 (Kluwer AP, Dordrecht, 2001), pp. 423–435 [hep-th/0008010]

Zac07. C. Zachos, *J Phys* **A40** (2007) F407–F412

ZC99. C. Zachos and T. Curtright, *Prog Theor Phys Suppl* **135** (1999) 244–258 [hep-th/9903254]

Zal03. K. Zalewski *Act Phys Pol* **B34** (2003) 3379–3388

Zdn06. P. Kaprálová-Žďánská, *Phys Rev* **A73** (2006) 064703

ZP94. W. Zurek and J. Paz, *Phys Rev Lett* **72** (1994) 2508; S. Habib, K. Shizume, and W. Zurek, *Phys Rev Lett* **80** (1998) 4361–4365;
W. Zurek, *Rev Mod Phys* **75** (2003) 715–775

Zu91. W. Zurek, *Physics Today* **44** (October 1991) 36; *Los Alamos Sci* **27** (2002) 2–25

# Index